A Potted Grammar

of

Natural Dialectic

Copyright © Michael Pitman 2017

2nd edition (coloured) 2019

ISBN 978-1-9999664-8-5

A catalogue record for this book is available from the British Library.

The right of Michael Pitman to be identified as the Author of this Work has been asserted by him in accordance with the Copyright, Designs and Patents Act 1988.

All Rights Reserved. Apart from any use expressly permitted under UK copyright law no part of this publication may be reproduced, stored in an alternative retrieval system to the one purchased or transmitted in any form or by any means electronic, mechanical, photocopying, recording or otherwise without the prior permission in writing of the Author.

Published by Merops Press

Website: www.scienceandphilosophy.co.uk

www.cosmicconnections.co.uk

www.scienceandthesoul.co.uk

www.michaelpitmanbooks.co.uk

Acknowledgements: Suzanne, Marianne, Emmanuel and Françoise.

Contents

Illustrations 6
Preface 7
Chapter 1:   Primary Assumptions 9
   Two Pillars: A Dialogue of Faith 11
Chapter 2:   Questions Arising 13
Chapter 3:   Models 14
   Mount Universe 15
   Concentric Rings 17
   Scales 18
Chapter 4:   Three Cosmic Fundamentals 21
Chapter 5:   Stacks 25
Chapter 6:   Models and Stacks 30
Chapter 7:   A Hierarchical Perspective 33
   Hierarchical, Triplex Construction of the Cosmic Pyramid 33
   The Cosmological Axis .... Error! Bookmark not defined.
Chapter 8:   Essence 38
   Source and Sink 41
   Voids 43
Chapter 9:   Extremities 47
   Subtendence 47
   Transcendence 50
Chapter 10:   Existence 54
   Causality 54
   Polarity/ Duality 57
Chapter 11:   Elements of the Basic Existential Dipole .. 63
   Upper Pole - Information 64
      (Sat) Potential Information 68
      (Raj) Active Information 68
      (Tam) Passive Information 68
   Lower Pole - Energy 69
      (Sat) Potential Energy 70

(Raj) Active Energy *70*
(Tam) Passive Energy *71*
Informed Energy *72*

Chapter 12: An Act of Creation **75**

Orderly Creation *76*
Law and Order *78*
Capability *80*
Purposely Down to Earth *80*
Getting Your Way - Pragmatics *80*
What Do You Mean? *80*
Information's Infrastructure - Code *82*
The Lowest, Physical Level *83*

Chapter 13: Triplex Psychology **85**

Triplex Mind *85*
The Neurological Delusion *86*
Does Brain Originate or Mediate? *87*
Consciousness *88*
Top-down Psychology *88*
First State of (Super-)Consciousness - *91*
   The Psychology of Transcendence *91*
Second State of Consciousness - *91*
   The Psychology of Waking Normalcy *91*
The Third State of (Sub-)Consciousness *93*
   The Psychology of Dreaming *93*
   The Psychology of Deep Sleep *94*
The Non-State of Consciousness - *94*
   The 'Psychology' of Physic *94*
Frozen Time *95*
Psychosomatic Linkage *96*
The Typical Mnemone *99*
H. archetypalis, the Image of Man *99*
H. electromagneticus *102*
Psychosomasis *104*
How Does the Connection Work? *106*
Whence did the Wired Side Emerge? *107*
Triplex Psychology Summarised *109*

Chapter 14: Triplex Physics **111**

(Sat) Potential Energy *111*
   Precondition *111*

    Lack of Cosmo-logic *113*
    Cosmo-logic *114*
    Can Mathematics Help Us? *116*
  Towards a Theory of Physic and Metaphysic *117*
  The Principles of a Unified Theory of Matter *118*
  (Raj) Active Energy *121*
  (Tam) Passive Energy *121*
  Space *121*
  Time *123*
  Alpha Points *125*
  Points Omega *127*
  The Matrix *128*

Chapter 15: Triplex Biology *129*

  (Sat) Potential Biology *129*
  (Raj) Active Biology *129*
  (Tam) Passive Biology *129*
  The Basis of Biology is Information *130*
  The Principles of a Unified Theory of Biology *131*
  The Central Executive is Homeostasis *134*
  Nuclear Super-Computing *135*
  Conceptual Biology *137*
  Darwin: Half Right, Wholly Wrong? *140*
  Chemical Evolution? *142*
  What's the Problem? *142*

Chapter 16: Truth, Appearance and Reality *146*

  Truth, Appearance and Reality *146*
  Two Value Systems *149*
  Rights and Wrongs *152*
  From Science to Conscience *153*
  Is There an Absolute Morality? *153*

Appendix 1 .... *155*

Glossary *157*

Index *177*

Illustrations

3.1	Mount Universe	15
3.2	Concentric Rings	18
3.3	Pivoted Existence	19
6.1	The Relationship Between Models and Stacks	30
6.2	Preliminary Model/ Stack for Biology	32
7.1	Cosmic Fundamentals and Their Ziggurat	33
7.2	Upper Sub-divisions of the Ziggurat	34
7.3	Lower Sub-divisions of the Ziggurat	35
7.4	Cosmological Bearings	37
8.1	The Diamond Capstone	39
8.2	Source and Sink are Zero	41
8.3	Zeroes	44
9.1	Subtendence and Transcendence	48
10.1	Duality within Unity	58
10.2	Primary and Secondary Duality	59
10.3	Primary Inversion	60
11.1	Conscio-Material Coordinates	63
11.2	Hierarchical Information	65
11.3	Three Tiers of Mount Universe	72
12.1	Microcosm of the Macrocosm 1	75
12.2	The Order of an Act of Creation	76
13.1	Microcosm of the Macrocosm 2	85
13.2	Essential Psychology	89
13.3	Mental Ziggurat	90
13.4	The Subconscious Sandwich	93
13.5	Grades of Man/ Dialectical Bio-classification	96
13.6	Wireless Man	98
13.7	H. archetypalis in Biology	99
14.1	Crystallisation of Principles	118
14.2	Physical Grades of Time	124
14.3	Alpha Answers	126
15.1	A Dialectical Plan of the Way Life Works	131
15.2	Neo-Darwinism - A Tabulation	140

Preface

Could our scientific outlook be lop-sided? Is balance needed? Is a fresh perspective possible?

This set of lectures is intended to abbreviate an explanation of what is, basically, a fresh and simple structure - **Natural Dialectic**.

Dialectic (or dialectical method) is to-fro discourse between opposing views in order to establish reasonable truth. Of course, there may exist several antagonists in a debate but, in the case of nature, it may be shown that only polarity or, at most, trinity, holds sway.

Natural, for its part, is often construed as a characteristic of any material item not devised by the mind or produced by the hand of man.

Natural Dialectic is one of different models mankind has used to try and understand the universe into which each one of us is, without asking, born. With roots deep in human thought and variously expressed at different times and places, the Dialectic's 'philosophical machine' reflects an oscillatory framework within which nature operates.

Any philosophical infrastructure, whether mathematical or verbal, is built of symbols, that is, of code or language. Language, involving a specific assignment and coherent arrangement of symbols, is the way that meaning is organised and information conveyed. Therefore, in order to understand, think of or communicate messages, it is helpful to have clearly grasped whatever particular grammar is being used.

In this case, does **Natural Dialectical Philosophy** accurately reflect the *modus operandi* of cosmos? And, as a whole, construe the grammar of polarity? It is, at least, the dynamic formulation within which the narrative of a parent, Science and the Soul (*SAS*), and consequent books is expressed.

You can, if you wish, exploit extensive Connections/ Endnotes (indicated on the Contents page) reaching back to these volumes in order to elaborate on queries raised by The New Look's considerable abbreviation. For easy reading, devices of **bold**, <u>underlines</u>, *italics* and red script simply draw attention to salient points.

Illustrations have for ease been coloured according to theme:

blue	universal/ cosmic
buff	informative/ developmental
violet	human extent

red-brown	unconscious region
green	biological
silver	morality

This book is, as regards its metaphysic, scrupulously, religiously non-religious.

Finally, after careful and perhaps exciting inspection you may better judge whether Natural Dialectic's grammar well interprets nature's text and thereby accurately reflects the logic of creation. So jump aboard and ride this streamlined 'thought machine'. Intellectual seat-belt fastened, we shall travel far and fast from here…

Chapter 1: Primary Assumptions

The first step taken on any journey is critical. If, for example, you intend to travel straight from London to Edinburgh and your first step is east or west you will not arrive.

In the case of cosmic world-view the first step is philosophical but also critical. Is your primary assumption, aimed towards full truth and understanding, correctly set or not? Which answer - materialism or holism - does the evidence best brace?

What are the basic axioms and corollaries of this pair?

Materialism's axiom[1] is that every object and event, including an origin of the universe and the nature of mind, are material alone; a few oblivious kinds of particles and forces compose all things. Moreover, cosmos issued out of nothing; therefore, beyond this realm of physics there is only void; and life is an inconsequent coincidence, electric flickers of illusion in a lifeless, dark eternity.

Although the universe appears to work by rules and to have been established in a very particular way, this appearance of order is in fact unplanned. Its invisible framework of regulation must have occurred by chance and, since inception, individual objects and events (called actualities) occur by chance as well.

Such axiom must apply to life. In this respect the **Primary Corollary of Materialism** states, by the neo-Darwinian theory of evolution, that life forms are the product of the chemical abiogenesis[2] of a first cell; and following that, by common descent, of a random generator (mutation) acted on by a filter called natural selection. Such evolution is an absolutely mindless, purposeless process. It is, from a materialistic perspective, a fact so that this *PCM* is a fundamental *mantra* of materialism.

Materialism has leapt to assume that what one cannot sensibly or physically test does not exist. No immaterial element exists. Is this an argument from ignorance or not? What precisely is it that believes an immaterial element lacks substance? *Isn't your own immaterial centre, receptive, thoughtful and creative mind, a fact?*

You are, of course, alive. You know full well, subjectively, life's consciousness - but is it proven physical? Your body's doubtless physical and you accept a cosmos made of matter; and if body is a special composition made of universal matter may we not holistically suggest

---

[1] Glossary under Primary Axiom of Materialism, *PAM*.
[2] Chapter 15: Chemical Evolution?; also *SAS* Chapters 20 and 21.

that, likewise, human form incorporates its special part of universal mind? To repeat, is not your mind, informant and informed, metaphysical? And, like your body, natural and part of something universal? Using naturalistic methodology[3] experimental science cannot prove life's central part, mind with subjective thought and conscious experience, is just a product of non-conscious particles and forces. *Holism, therefore, simply adds immaterial, as a second fundamental cosmic ingredient, to material. Or, conversely, it adds material to immaterial.* **Thence follows this philosophy's impregnable validity.**

Natural Dialectic, an expression of it, is holistic. This means that, in addition to materialism's single cosmic component, matter, there is added an antagonist.

In this view (the opposite of reductionism) a whole is greater than the sum of its physical parts; the extra, metaphysical ingredient is identified by Natural Dialectic as ***information***. Information, which occurs in active and passive modes,[4] implies the purposeful design, development and arrangement of contingent parts in a working system. Such a system embodies the Principal Idea and subsequent ideas that drive a coherent program or machine. In computer terms, you could say that a Main Routine controls sufficient subroutines.

This simple, holistic premise is powerful to the extent that it relates all physical and psychological phenomena that humans can appreciate. Hence arise two propositions central to both the book and the cosmos it describes.

Holism's axiom[5] is that realistic comprehension of the world includes *two* primary components - immaterial and material or, as obvious to everyone, mind and matter.

Is there really any difference? Isn't consciousness unconsciousness? Matter mind? Isn't a material brain the same as, or at least the generator of, your mind? Aren't you your body? It is made of cells, cells are made of chemicals, chemicals of atoms and atoms aren't alive. If atoms, molecules and cells aren't then your body isn't. Alive is not the same as lifeless. It might be a marvellous machine but it is not alive. *So who are you? Are you alive or dead?* **It follows that a scientific world-view that does not profoundly and completely come to terms with the nature of conscious mind can have no serious pretension of wholeness.**

What, moreover, was before the world began? What is the nature of such nothingness whose logic or its lack substantiates creation, chemicals and bodies? Creation as a whole, the science in us feels, is

---

[3] see Glossary
[4] *SAS* Chapters 6 and 17: Informant and Informed Domains.
[5] Glossary under Primary Axiom of Natural Dialectic, *PAND*.

'logical'. **The second proposition is, because our understanding reasonably reflects it, that existence as a whole *is* 'logical'.** Such logic can be intellectually expressed in various ways. Physics chooses mathematics but we'll trace cosmic contours with another kind of symbol, one that probes where numbers cannot pass - words.

Such axiom must apply to life. In this respect the **Primary Corollary of Natural Dialectic (*PCND*)** states that the origin of irreducible, biological complexity is not an accumulation of 'lucky' accidents constrained by natural law and death. Forms of life are conceptual; they are, like any creation of mind, the product of purpose. Such assertion is, in the face of materialism, absolute anathema. Yet, if materialism's first axiom is incomplete then every step that follows will lead further from original truth. *An axiom that discounts the force of information may well be largely incomplete.*

Two Pillars: A Dialogue of Faith

These apparently antagonistic axioms amount to assumptions on which theistic and atheistic creeds depend. **Both assertions are philosophical; neither is a scientific one.**

We shall find (Chapters 3, 4 and 11) that energetic *and* informative causations are the way the world proceeds. Of these, materialism tracks the physical and energetic: holism also tracks the psychological, informative domain of mind. The former is pitched outwards with an interest in material creation and the latter inward towards the only known source of creativity. Such world-views are closely allied to tendencies of concentration.[6] On the one hand, there is concentration 'down and out' through the senses to everything external to mind; this sensible world includes the biological body of its thinker. On the other, there is concentration 'up and in' towards consideration, working out, understanding principles and, eventually, comprehension of the source of conscious mind itself.

***Top-down* and *bottom-up* are designations that describe these two directions but also fundamental vectors[7] embedded in the way the existential dipole works.** As you will come to understand, the Natural Dialectic of Polarity[8] consistently contrasts these anti-parallel vectors of comprehension.

To understand and to make (or do) are anti-parallel modes of behaviour.[9] Each involves its main direction of concentration; and each

---

[6] *SAS* Chapter 0: Opposite Directions of Mind.
[7] see chapters 3-7; also *SAS* Chapters 0-3 and Index: anti-parallels and cosmic fundamentals.; and *AMA?* Glossary: anti-parallel perspectives.
[8] see Chapters 10 and 11 as opposed to the Natural Dialectic of Singularity (Chapters 8 and 9: Transcendence).
[9] *fig.* 13.2 Essential Psychology.

direction is reflected in perspective, method of study and, as mentioned above, world-view.

Bottom-up is taken as the empirical method of a humble student who, from child-like ignorance, starts from knowing nothing. Such lack of preconception marks a strength of scientific method. Its student learns by experiment and experience; and at the same time is guided, *top-down*, by the relative certainties of previously acquired knowledge and by a higher authority, a top-notch teacher.

Top-down implies you've got the information that you need; from 'on high', it is a system maker's expert point of view.[10]

This pair constitutes the anti-parallels of knowledge. So different are the 'world-views' derived from their 'opposing' perspectives that, whenever the counterpoint of this contrasting Dialectic is expressed, each party is habitually italicised.

Time for a summary.

We repeat the primary assertion - mind and matter are two separate elements.

But, you respond, materialism's primary axiom is that there is only one. Precisely so. That is non-materialism's simple null hypothesis. **But let us at the outset be completely clear - both assertions are philosophical; neither is a scientific one.** Nor can material science ever prove holism's metaphysic, based on mind and information, untrue. *Materialism, like holism, is a philosophical and not a scientific posture.*

Moreover, if materialism's arguments and evidence are found weak and wanting then it is right to elaborate a logical alternative.[11]

---

[10] For example, *top-down* computer programs involve a cascade of subroutines that issue, according to conditional commands, from a top or master routine. Such algorithmic control embodies both purposeful idea, means of execution and, in its end-product, fulfilment. *Top-down* programs, although found at the codified root of bio-formation, *never* happen accidentally. See Index: computation and hierarchy; also *SAS* Chapters 6, 19, 24, 25 and Index: bio-logic and computation.

[11] *SAS, A&E, AMA?* and several books of research notes each add to demonstrate this case. *PGND* sets out to clarify the infrastructure, Natural Dialectic, into which facts may be fed.

Chapter 2: Questions Arising

If the holistic axiom that mind and matter are two different kinds of element is true the logic of this book in its entirety is unassailable. **This logic is drawn, over the chapters, into a self-consistent, polar model of creation; and, paradoxically transcending this polarity, into consideration of a causal singularity.**

Of course, such axiom exacts a toll. A fee needs to be fully paid. Costs that need, like stinging nettles, to be grasped, include:

(i) the nature of consciousness[12], sub-consciousness[13] and non-consciousness.[14]

(ii) whether individual mind can exist independent of a body and, if so, the nature of its 'entry', attachment, 'exit' and disembodied condition.[15]

(iii) the interactive relationship of individual mind with body; the nature of any *PSI* (psychosomatic border or, perhaps, quantum linkage) between mind and matter.[16]

(iv) the mechanism by which universal mind, if such exists, might inform non-conscious forces, particles and gross phenomena; the origin of physical constants and patterns of behaviour, that is, the laws of nature.[17]

(v) the nature of physical and biological prototypes, homologies or, if any, archetypes.[18]

(vi) the question whether biology is informed by chance and aimless natural law or by design in accordance with such law; a wholesale reappraisal of the neo-Darwinian theory of evolution.[19]

To address these issues we'll need relevant knowledge drawn from the broad physical disciplines of biology, chemistry and physics; and, beyond materialism, from the disciplines of psychology, information theory and philosophy. How, put simply, can this be?

---

[12] *SAS* Chapters 5, 6, 13-14.
[13] *SAS* Chapters 15 and 16.
[14] *SAS* Chapters 7-12.
[15] *SAS* Chapters 13 and 18.
[16] *SAS* Chapters 6 and 15-17.
[17] *SAS* Chapters 5-12, 15, 16.
[18] *SAS* Chapters 7, 8, 15-17, 19, 22 and 24.
[19] *SAS* Chapters 5, 6, and 19-25.

Chapter 3: Models

Scientific and non-scientific concepts need explanatory metaphors. Metaphors and models are conceptual hooks. What's invisible or immaterial is hung on images (or, if you like, imaginations). These can mislead, approximate or usher towards a clearer model and a closer truth. For example, science has employed successive models of the atom as it edges closer to atomic actuality.

In short, mind needs models like a handle so that it can grasp abstraction. If you accept that metaphysic might exist what metaphors or models can proponents of holistic outlook use? Within what philosophical sort of frame might we best arrange our ideas of the world?

Natural Dialectic is a 'philosophical machine' whose operation is described over the next four chapters. Basically, it involves a well-known dynamic. For example, play of opposites underwrites the Ancient Chinese philosophy using *yin* and *yang*. The Classical Greek mode of discourse commonly employs the particles μέν ('men', meaning 'on the one hand') and, as its opposing or alternative force, δέ ('de' as in debt meaning 'on the other'). Manichean and most other religious traditions see the world in terms of opposites such as 'light and dark', 'good/ bad' and so on.[20] And, of course, our entire information age is based upon an on-off switch composed of binary digits. Fundamentally computers run on 1 and 0.

So Natural Dialectic's **ABC simply asserts that to-fro, binary logic is the natural way all things work; and, furthermore, underlies the way our polar cosmos is, as a whole, constructed.** Equally, it asserts that at the heart of such polar infrastructure lies balance, neutrality or potential; and that such balance represents pivotal unity at the heart of polar action. **In this way, trinity better describes creation than duality.**

Put this another way. **Dynamic cosmos oscillates within a pair of fundamental poles (mind and matter; see** *fig.* **10.2); and may, at points of balance, rest from flux.**

Mobile, to-fro relativity and equilibrium. Equilibration is a cosmic operation too. Balance and equation rest beneath the restless show. Two-in-one; one-in-two. Unity, duality *and* trinity. In the next chapter we'll meet the fundamental three and if, in fact, the world embodies them you must judge how closely Natural Dialectic comes to frame its truths. In trinity it well reflects relationships between all kinds of complementary pair; and, beyond them, links each couple with its pivotal control, that is,

---

[20] see further in Appendix 1.

its mean or government. **Indeed, Natural Dialectic is a theory of principle that, setting up its framework of polarity, makes possible a universal and yet self-consistent description of the natural order.**

It uses three models to illustrate this order. These are **Mount Universe**, the radiation of **Concentric Circles** and a **Pair of Scales** (or a Balance). The framework within which its flows are registered is called a dialectical stack. The operation of these stacks will be explained in Chapter 5.

Mount Universe

Mount Universe, in every aspect of its whole, involves gradients. *These run from top to bottom, high to low or source to sink.* **They involve, in holistic terms, gradients of energy and information**. The former ranges from high, subtle and free down to gross and locked in static mass; and the latter from highly concentrated, creative focus down to uncreative, automatic reflex, locked repetition or at sink, oblivion.

We go further and propose creation shows them as inverse proportions of each other on a conscio-material scale. This spectral view of cosmos is, as we'll see, reflected well in one of many diverse forms - your own.

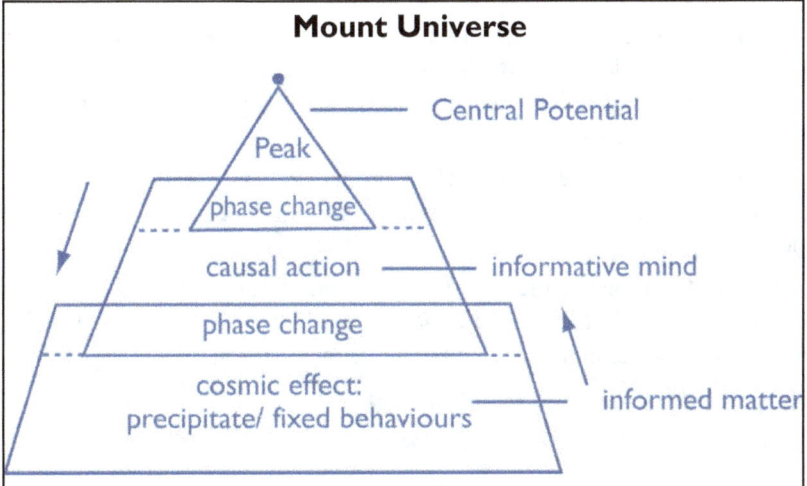

A cone or, squarer, pyramid describes 'static' hierarchy. A useful representation is the stepped pyramid, also called a **ziggurat**. In this case each step of a **ziggurat** stands clearly for a phase, level or stage; and the apex of its capstone, a point that points beyond the finite grades below, implies peak infinity. **This capstone is the highest point and source of what we call Mount Universe.**

fig. 3.1

Thus the *first* image to represent the gradient of creation is **Mount Universe** or, in the conceptual geometry of a cone or ziggurat, **The Cosmic Pyramid**. Levels of mind and matter display the rainbow-like continuity of a spectrum but states of matter also show a discontinuous, phased aspect to their energetic levels; thus a smooth pyramid becomes, converted into discontinuous levels, a stepped structure called a **ziggurat**. The model is, rather than energetic, 'heavy'. It is an inertial, solid picture of existence. Its grades are not smooth; the phases change (as in the case of sleep/ waking or solid/ liquid and gas) in 'jumps'. Welcome, then, unto the Dialectic's tetrahedral ziggurat. In the tiers of a pyramid, you obtain a clear picture of both hierarchy and duality.

base	*peak*
low	*high*
sink	*source*

You can also note rise and fall by writing a *triplex* 'stack':

↓ *down/ descent*	Peak	*up/ ascent* ↑
sink	Source	flow
negative	Neutral	positive
impotent equilibrium	Equilibrium	action zone

Source above sink. Hierarchy always runs from source to sink; it falls between these poles. It forms a gradient of creativity or power. **Indeed, you can think of intelligence, information, energy or bulk materials in terms of concentration gradients.**[21]

Think, for example, of the concentration of information found in a general principle. Principles underwrite ideas which lead to specific action and results.

Physically, think of the energetic concentrate of a star and its radiant diffusion of light into cold, dark space. Or, if you like, consider electrical or chemical diffusion down a slope from concentration, source or origin.

↓ *end-product*	Potential	*creativity* ↑
effect	Cause	stimulant
end	Origin	process
black	White	spectrum

Prior to action any single thing is, almost egg-like, in 'potential equilibrium'; such stillness, readiness or poise is full of certain possibilities. There follows (if some fluid circumstance permits) a spread of chemicals, a reaction, flow of current pole to pole or growth of seed to

---

[21] The notion of scale is introduced; so also, by extension, the notion of a universal conscio-material gradient.

adult form. An event rolls to its close; exhaustion is potential's opposite, inertial kind of equilibrium.

In short, at the very start of motion is poise; the source of action is a 'spark' or 'tipping point'. Thus the equilibrium of potential discharges into impotence; the latter's flat equilibrium constitutes a polar reverse.[22]

You can, by correctly introducing energy, recharge a battery. How, up against its outward tide, could a cosmos be recycled back to source?[23]

At any rate, in this view potential, whether in the form of energy or information,[24] is a source. It is replete with charge that is by action discharged into impotence, its sink.

You can antithesize material and immaterial elements:

material	*immaterial*
physic	*metaphysic*
energy	*information*
informed	*informant*

And then objective and subjective ones:

objective	*subjective*
outside/ outer	*inside/ inner*
matter	*mind*

In this way you may arrive at beginning to understand the idea of a gradient inhabiting not only the physical but metaphysical cosmos; you might grasp the 'Jacob's ladder' of a spectrum that, from source to sink, describes creation.

↓ *non-conscious matter*	*Potential*	*forms of mind* ↑
passive	*Precedent*	*active*
sub-state	*Super-State*	*action zones*

We have already reached fresh territory. The habitat is unfamiliar and, because all can't be said at once, some connections made in stacks may seem abrupt or even disconnected. Hang on. We'll soon understand the sense of Natural Dialectic's ways.

Concentric Rings.

Let us now turn to the <u>second</u>, energetic model of the cosmos. It is one of <u>Concentric Rings</u> (or, three-dimensionally, concentric spheres) losing power as they recede from Source.

---

[22] see also Chapter 8: Neutralities and Chapter 10: Polar Inversion.

[23] *SAS* Chapters 5: Top Teleology; 12: Points Omega; and 18: Immortalities.

[24] see Chapters 11 and 12: The Basic Existential Dipole.

Concentric Rings

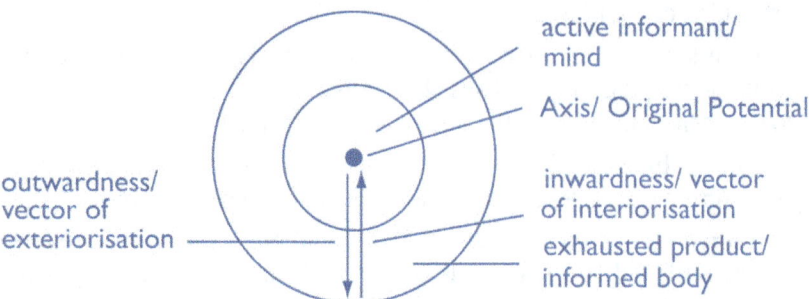

This figure of concentric rings includes antagonistic vectors to and from its Central Source. Is this not, from generative heart to petals, how the patterns of our cosmos all concentrically flower?

fig. 3.2

Images of radiant energy, familiar to physicists and mystics alike, include Bell, Light or Pool. Surely you have dropped a pebble in a pool? The action describes a central source, a projection of power and a gradient of vibrant energy - just as light dims with distance or the ripples on the pool faded.

↓ *outward vector*	*Axis/ Hub*	*inward vector* ↑
loss (of info. or energy)	*Central Origin*	*gain*
entropy	*Constancy*	*negentropy*

In an outward, materialising direction, involving entropy, clarity of an original message becomes more or less quickly garbled. It loses sense. And fluid waves diffuse or else are locked by cooling into crystalline precipitate.

In short, potential energy once activated flows through kinetic down to an exhausted level. **Such a triplex state of motion describes, we find in the next chapter, a fundamental characteristic of the universe.** It marks the gradient of every object and event.

Scales

Thirdly, you can think of cosmos as a <u>Pair of Scales</u>.

↓ *down/ descent*	*Point of Balance*	*up/ ascent* ↑
heavier	*Pivot*	*lighter*
minus	*Zero*	*plus*
sink	*Source*	*action/ flow*

Two pans of a kitchen balance oscillate about their starting-point. Opposing vectors swing around a point of balance. The motion of creation weighs upon a fulcrum of poise. Such equilibrium amounts, as we shall see, to potential from which action springs. You can, by equation, weigh the cosmos in a way, at base, as simple as the scale you have in mind.

In this useful picture ups and downs swing round a pivot by degrees.

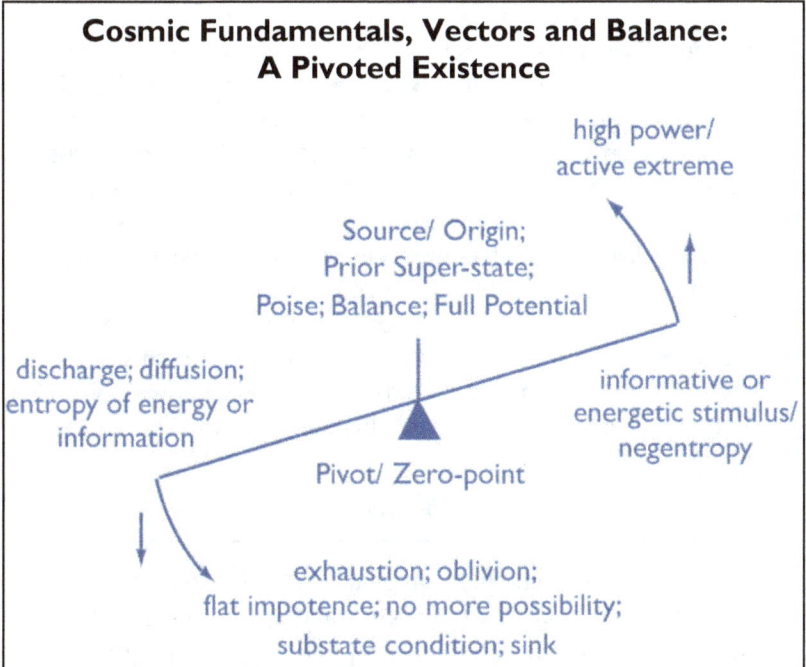

Imagine creation as a scale in which every movement is counter-balanced. The translation of the order of such creation - from potent super-state through phases of action down to sub-state impotence - is also rephrased in terms of spectrum, ziggurat and, as we'll shortly see (following Chapter and *fig.* 6.1) three cosmic fundamentals. Each fundamental is predominant at one of three major levels of creation.

Fig. **3.3 makes clear that, in description of swing around a pivot, potential *is* that pivot.** Written dialectically 'above' or preceding action, it is the point whence wobbling changes in the two domains of mind and matter start; it is the point of poise round which all swing begins and of equilibration when a balance is assumed again.

The cosmic scale integrates a pivotal, balancing factor (*Essence*) with two antagonistic vectors of existence. In this, the perpetual changes of creation

> **are seen as myriad adjustments against the disturbance of balance.** Up, down - nature is a scale whose beam forever wobbles in equilibration round its various centres; and, on the large scale, one could see such existential instability equilibrating round a Central Pivot, an Essential Point of Balance. Stable Axis represents Perfection.
>
> Existence, as opposed to Essence, is a field of ceaseless change. Creation's wheel is, therefore, one of relative uncertainty and instability. **Such eccentric instability is forever perfectly imperfect; perfect imperfection shows as the perpetual motion of continual changes at local times and spaces.** Existence thus amounts to a swinging but self-regulating balance; it amounts to the sum of myriad individual actions and reactions each of which always shows proportions of two vectors pivoted around the poise of a third non-vector. Such orderly procession by equilibration is sometimes referred to as the '*karma* drama', physical and psychological fields of *karma* or equations of creation. Physics and chemistry also measure transformations by equation.
>
> **fig. 3.3**

From now on we shall study pivoted, vectored existence. The three event vectors are balance (the nul vector) and up (↑) and (↓) down.

In pursuit of what this means we turn to consider Natural Dialectic's cosmic fundamentals. Let's see what they mean and how the stacks they govern work.

Chapter 4: Three Cosmic Fundamentals

Is three nature's lucky number? It is now obvious from physics, event vectors and Natural Dialectic that we are not dealing with dualities but *trinities*. Take a scale's three phases - balanced and swinging upward or downward. **The actions that these words symbolise find various expression in each object and event.** More subtly, they reflect the *qualities* intrinsic to creation. They compose three fundamental operating principles whose permutations are expressed, in varying degree, as *tendencies* both psychological and physical. **We call them cosmic fundamentals.**[25] Ups-and-downs and peaceful poise; ins-and-outs and equilibria; existence is an ever-changing play, a field of pulsing tensions that evinces, as already mentioned, trinity. *If the triplex is naturally fundamental it is also fundamental to the operation of this Dialectic; but while the orient has long worked with these immaterial radicals western minds have not.*

↓ *yin*   *Tao*   *yang* ↑

You might be familiar with this far-eastern, Taoist set and its associations.[26] Therefore, I suggest using an equivalent, equally ancient abstraction whose purity is, because of occidental unfamiliarity, less stained by prejudice or shadowed by prior connotations. Let's employ (despite, perhaps, conservative resistance) the following as-yet faceless trio.

↓ *tam*   *Sat*   *raj* ↑

***Sat*, *raj* and *tam*.** Those interested in dietary cooking might have already picked up on their categories. For example, *sat* food is fresh, light and includes fruit, vegetable and cereal products. *Raj* food stimulates; it is hot, promotes physical activity and includes spices, curries etc. A *tam* ingredient is stale or heavy; it includes meat and alcohol. You get the flavour. *In fact, the triplet (each member called in Sanskrit a guna or a thread) is much less prosaic and more connective.* **It is all-pervasive to the extent of describing the three major tiers of Mount Universe itself.**

↓ *tam*	*Sat*	*raj* ↑
informed	*Potential Informant*	*informant*
subconsciousness	*Concentrate of Consciousness*	*consciousness*
passive mind	*Source of Mind*	*active mind*

---

[25] *SAS* Chapter 1.
[26] further research would also indicate an extensive association between Natural Dialectic and Sankhya, Vedanta and Buddhist doctrines. Check also Appendix 1.

A simple link from <u>psychology</u> describes the three basic conditions of information as (sat) potential informant (concentrate of consciousness) prior to (raj) active mind (both informant and informed) and (tam) passive, dormant or sub-conscious mind. Of course, *passive information* may be *carried* by material forms or memory; but *active information* is the character of consciousness.

Psychology's the study of an immaterial information-centre - mind. (*Sat*) **consciousness is the potential for** (*raj*) **active thought and creativity; exhaustion falls into** (*tam*) **passive sleep, that is, a subconscious zone.** And our normal waking zone clearly recognises 'higher' or 'lower' levels of, for example, intelligence, moral and aesthetic quality, rationality and happiness.

↓ *tam*	*Sat*	*raj* ↑
informed	*Archetype*	*informant*
matter-in-practice	*Principle*	*matter-in-principle*
gross appearance	*Potential Matter*	*subtle quanta*
locked energy	*Precondition*	*free energy*
exhaustion	*Latent Behaviour*	*stimulus*

A simple link from <u>physics</u> describes the three basic conditions of energy as (sat) potential prior to (raj) action and (tam) exhaustion.

Physics is the study of non-conscious, energetic transformations.

↓ *tam*	*Sat*	*raj* ↑
output	*Informative Archetype*	*coordinated program*
hardware	*Systems Architecture*	*software*
informed body	*Conceptual Integrity*	*informant code*
associated machinery	*Balance Point/ Mean*	*orderly operations*
structure	*Reason*	*function*

Thirdly, *a simple link from <u>biology</u> describes the three basic conditions of life on earth as (sat) informative archetype, (raj) metabolism including a code carrier called DNA and (tam) finished product called developing or adult body.*

The biology department studies a fusion of psychological and physical elements. It studies soft machines called bodies; also instinct and behaviour. The potential for material bio-operations is a plan subtly codified in the language of molecular *DNA*; this genotype is, in turn, expressed as a gross physical result called a phenotype - the temporary and dynamic appearance known as an organism's body. Potential form; codified transfer; expression incorporated in a product such as yourself.

The following chapters elaborate, in Dialectical terms, the hierarchical meaning of these basic linkages. Start, process, outcome; poise, excitement, exhaustion; balance (*sat*), up (*raj*) and down (*tam*) compose the triplex nature of creation.[27]

How does such triplex modulation work? Natural Dialectic develops a very simple, columnar structure to illustrate, in an orderly and connected way, the myriad expressions of just three basic abstracts. These abstracts are not any particular phenomenon or its components but exist, proportionally, *in* every one.

It is, therefore, a distinct advantage that the terms chosen by Natural Dialectic to label these fundamentals - **sat**, **raj** *and* **tam** *- are fresh, unaccustomed and uncoloured by particular preconception or prejudice.* What does each stand for?

↓	*tam*	*Sat*	*raj* ↑
	from nuclear	*Nuclear*	*towards nuclear*
	to periphery	*Central Origin*	*from periphery*
	fall	*Peak*	*rise*
	effect	*Potential*	*stimulus/ cause*

<u>Sat</u>, the Essential or Central Quality, comes first. *Sat* means, in Sanskrit, true. Its 'truth quality' is in the paradoxical department. It represents the 'deep principle' of unity that underlies duality. It is the *Tao* from which *yin* and *yang* divide. *Sat represents the whole. Including both sides of an equation it is, as such, the 'equals sign', the factor of equilibration.* Thus it can be viewed not only as fulcrum but as the centre or apex of any systematic creation. It is the point of origin and originality.

The essence of this principal is consciousness or, as we understand it, life. *Sat* is, equally, potential's quality; it involves prior, implicit capacity whose possibilities, when reduced, will have informed and defined an explicit, actual end-result. It involves, in other words, principle, intelligence and law. This applies to the way you work.[28] It also applies as much to the universe with respect to the minuscule objects/ events from which it is built up. Thus its character is source not sink, start not finish of a process of creation. In terms of expression call it precondition; in terms of initiation call its primal-unity-encompassing-polarity an egg, the cosmic egg; and in terms of physical effect understand metaphysical cause. **Informative potential, plan, precedes material behaviour.**

Sat is high-level, at the axis or top pole of things. *Its purity is vested in a super-state, that is, a Transcendent Concentrate of Information at the Apex of Mount Universe.* In terms of local, energetic concentration call it 'sun'. As such it blazes for all satellites; the brilliance of its influence is variously reflected by those influenced. *Sat* shines, physically, with

---

[27] see Chapter 7: Hierarchical Perspective and Chapters 13, 14 and 15 for the three abovementioned scientific disciplines.

[28] see Chapter 16; also *SAS* Chapters 14, 16, 17 and 19.

energetic light and heat. Psychologically, informative illumination imbues, in varying degree, the qualities of comprehension, knowledge and wisdom; and infuses the information implicit in an ideal, purpose or construction. In truth it is the source and principle of order, coordination and coherence - the superintendent principle. It both resolves the tension between poles and is, as zero-point, the axis of balance around which fluctuations swing. Its immaterial equilibrium gives rise to the two other swaying, mobile tendencies (of *raj* and *tam*) and, in this sense, it is also their transcendent, causal point of origin.

Raj and tam, under sat, are the antagonistic fundamentals.

These two are inversions of each other. *Together they break the absolute symmetry of pre-formation; they act to express any potential.* Pole and anti-pole, they represent opposites, ranges and relativities. **Scales tip, pendulum swings, the world oscillates. Perfectly, imperfect, always out of Balance, cosmos cycles.** These cycles are its beats, its life-beats. Motion downwards, irregular, spluttering or losing rhythm is *tam*; but, pushing forward with regular and well-timed swing, *raj* is on the vibrant up-beat.

<u>Raj</u> (action ↑) is represented on the right-hand column of Natural Dialectic.

Action, like motion, is one thing but its vector is another. *Raj* ascends and its climb, *as the vector of levity and of negentropy,* becomes increasingly buoyant. It returns a pan's descent back towards equilibrium, equilibrates or moves towards balance. It 'ascends', in other words, towards a culmination of relativity in Absolution.[29]

<u>Tam</u>, on the left-hand column, is the vector of 'fall'. It is the negative, passive, inertial quality. It is the (↓) materialising tendency.

Tam drags. Tam exhausts. Its negative arrow subtends towards base (called 'sink'); its 'purity' is represented by the sub-state pole. As opposed to *raj*, an activator, *tam* contains and resists. It represents the medium which energy informs, the 'solidity' 'fluidity' has patterned. *Tam* rigidifies. *It is the vector of manifestation expressed as gravity, binding energy and crystallisation.* Mind's a multiplier; from principle to practice, from theme to detail[30] it's a differentiator. Similarly, as gas drops to solid state so *tam* vector precipitates out through the regions of the universe. At the same time as it solidifies it divides, at base level, into fixed and separate details; subtle atomic principle is expressed as coarse aggregate; *tam*'s restraining action ends with an inertial equilibrium. In descent there manifest increasingly extreme forms of properties such as impotence, massiveness, unawareness and destruction. In terms of process *tam* represents descent towards finality.

---

[29] see Chapter 7; also *SAS* Chapters 3, 4, Glossary and Index.
[30] see, for example, *fig.* 6.1.

Chapter 5: Stacks

Natural Dialectic is expressed in '*stacks*'. You are already familiar with a few of these verbal diagrams. They constitute a shortcut, quasi-pictorial way to simplify a complex world. Indeed, in conjunction with the cosmic fundamentals they form a triplex archetype from whose substance the character of all complexities (including the various 'fundamental' laws of chemistry and physics) ramify. Dialectical stacks of opposites are not, however, mathematical. Nor, although conceptually simple, are they trivial or wrong!

negative   *positive*

Let's start with a simpler 'duplex' presentation of the case. **This is because Dialectical stacks are a columnar expression of polarity**. Each **member** of a stack (e.g. negative *and* positive) consists of a pair of polar 'anchor-points'; and each **element** of the pair (negative *or* positive) is arranged according its fundamental characteristic viz. its tendency to (*raj* ↑) rise or (↓ *tam*) fall.

In addition to this member we could, for example, stack:

negative   *positive*
fall   *rise*
exhaustion   *stimulus*

It is immediately obvious that members of a stack[31] are not synonymous. But they equate in fundamental character; they represent equivalence in terms of cosmic fundamentals.

Also, each member implies a scale or dynamic range that runs between its elements, that is, its 'paired opposition' or 'complementary covalency'. For example, you can have 'more or less negative'; you may suffer a 'greater or lesser' fall. *As a scale of greys runs between extreme black and white, so it is implied that a scale permits oscillation of values between any pair of elements.*

On any scale motion runs down (↓) and (↑) *vice versa*. Or you could visualise a horizontal spectrum with your arrows oscillating (←) down or (→) up. For example:

↓ dark ←   → light ↑

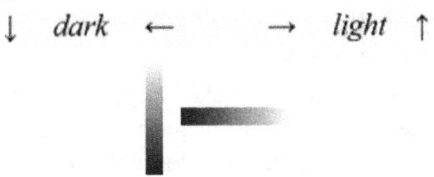

---

[31] a stack is a set (or pile) of members; see also Glossary: dialectical stack; also *SAS* Chapter 1: Natural Dialectic.

Do you call an electron's charge positive or negative? Edison flipped a coin and called it negative; but, because an electron is stimulant and a proton massive/ gravitational in character, the Dialectical preference would be unconventional viz. electron positive and proton negative.

Of vertical and horizontal visualisations one is superfluous so, with a similar flip of coin, Dialectical convention prefers the vertical, up-down representation. A single, non-repetitious first-line representation elegantly indicates the direction of elemental vectors for the whole stack. For this reason you will not see any more horizontal arrows. Instead, to indicate the presence of such implicit but dynamic scale we add anti-parallel arrows to the first member of a stack:

↓ *dark*   *light* ↑

or

↓ *negative*   *positive* ↑
fall   *rise*
towards sink   *towards source*

The vector of output grades from source to sink, action to exhaustion or, if you like, cause to effect. Conversely, the vector of input runs the other way. Of course, time's street is generally one-way but local two-way traffic is sustained by temporary, energetic inputs. Only at quantum and universal levels does perpetual motion appear to substantiate such basic particles as electron and proton and to instantiate an expanding universe.

↓ *tam*   *raj* ↑
negative   *positive*
divisive/ polarising   *depolarising/ unifying*
exhaustion   *flux*
lock-up   *freedom*
materialisation   *dematerialisation*
creation   *dissolution*

Creation: dissolution. The connections of membership in this stack will become clear. For now let's remember that an important way of understanding Dialectical dynamic is that materialising currents flow down (↓←) from right to left; and, conversely, the vector of dematerialisation rises (→↑) from left to right. These two gradients are the way, at root, that cosmos works.

However, a set (or 'duplex' stack) of dipoles is clearly not the full story. For example, positive and negative are both antithesized by neutrality. In physics, charged particles can be created from neutral (uncharged) light and, conversely, annihilation of matter by anti-matter yields light. And a neutron is a combination of charged particles. In this case a triplex or tri-logical stack might read:

↓ tam	Sat	raj ↑
(-) negative	Neutral	positive (+)
down/ descent	Equilibrium	up/ ascent

or (from Chapter 3, A Pair of Scales)

↓ down/ descent	Point of Balance	up/ ascent ↑
heavier	Pivot	lighter
minus	Zero	plus
sink	Source	action/ flow

However, it is sometimes more convenient to write a non-vectored di-logical stack and then polarise its neutrality into a second (*raj/ tam*) vectored and also di-logical membership. What is meant by this? How is it done?

polarity	Neutrality
duality	Unity
motion	Inaction
existence	Essence

The first, top stack of the pair disposes Essence and Essential Characteristics on the right against those of existence on the left.

Such a binary stack is called Primary, Essential or Central Dialectic. It is indicated by writing the right-hand column with a capital letter.

A Primary Stack sets (*Sat*) Unity against (↓ *tam/ raj* ↑) duality; or, if you like, it sets qualities of motion against Inaction or relativity against Absolution.

The stack above sets, fundamentally, the Equilibrium of Essence against ever-moving, ever-changing, localised existence. Although binary, such Dialectic involves a neutral, generalised expression of its existential component. And because (*Sat*) elements involve no movement, no arrows are attached. The stack is *non-vectored*. Also, since it involves qualities of Essence, Primary Dialectic is placed, as the superior table, above its polar, existential counterpart.

↓ existence ↑   Essence

Duality, motion and existence in general *are* vectored. These vectors represent direction of the (↑) *raj* and (↓) *tam* cosmic fundamentals. *Indeed, stacks represent the operation of such fundamentals*. Although the anti-parallel arrows of their dynamic are shown in the member of this Primary Stack, in practice they are superfluous and thus, by convention, dropped. They will occur in the secondary stack, always attached below the Main or Primary Routine, whose duality will polarise the left-hand column's elements.

For example, *polarity* itself is split into *positive* and *negative* components. And if, for example, you take 'motion' from the primary stack above and make a secondary, polarising split:

↓ *tam*	*raj* ↑
inertia/ drag	*stimulus*
inactive immobility	*active mobility*

In other words, you may set unity against duality. Duality, however, implies polarity. **Such polar component is expressed in the lower, vectored so-called <u>secondary, existential or polar dialectic</u>.** Thus **secondary, existential stacks** (*written exclusively in lower case*) represent the various kinds of polarity from which the changeful web of existence is composed.

issue	*Source*
motion	*Inaction*
↓ *down*	*up* ↑
sink/ immobility	*mobility*

You may have noted that, paradoxically, 'inaction' is action's before *and* after. How might this be? It is, indeed, an interesting and important question how the state occurs in both right-hand Primary and left-hand secondary stacks. The issue will be dealt with in Chapter 7.

In brief, inferior, existential stacks represent the various kinds of polarity from which the changeful web of existence is composed; and each pair of so-called 'polar anchor-points' implies a scale or dynamic range that runs between 'paired opposition' or 'complementary covalency'. Nature's scale is, basically, one of relativity.

We have seen that gradients (scales, slopes, degrees - call them what you will) may be illustrated as smooth (in the case of spectrum or spectral cone) or phased (ziggurat). The gradients themselves may represent, at root, level of informative or energetic focus, that is, concentration.

In quantum (fundamental) terms the physical cosmos is energy-based; and it is vibrant. Since vibrant oscillation is a wave you might think of changes in scale as motions like those registered on a sound meter, digital display or, as waveforms, on an oscilloscope. Physics is replete with equations that represent oscillatory calculus; such calculus, wheeling up and down between a scale's extremes, is Dialectically vectored. And what is a musician's score but the orderly arrangement of vibration?

At this stage let's juxtapose a Primary and secondary stack in the full and normal way:

Primary Dialectic:

To reiterate: *The first group disposes Essence and (Sat) Essential Characteristics, capitalised and on the right, against those of existence on the left.* **Such a binary stack is called Primary, Essential or Central Dialectic.**

tam/ raj	*Sat*
existence	*Essence*
polarity	*Neutrality*
expression	*Potential*
limitation	*Infinity*
duality	*Unity*
relativity	*Absolution*
motion	*Balance*
something	*Nothing*
(3)/ (2)	*(1)*

Secondary Dialectic:

↓ *tam*	*raj* ↑
fall/ down	*rise/ up*
negative	*positive*
division/ multiplication	*unification*
isolation	*connection*
drag	*stimulus*
(3)	*(2)*

The lower stack is called secondary, existential or peripheral dialectic. It is placed below the 'superior' Primary.

Natural Dialectic asserts that the to-fro, binary logic these stacks help to illustrate is the way that all things work; and is, furthermore, the way our polar cosmos is constructed. Such radical infrastructure involves a trinity of fundamental qualities; and it hints that, if there is an Absolute Reality, it will not be found in existence. Unity transcends duality.

In summary, stacks of members and their elements comprise a systematic way of loading opposites into a triplex structure based on cosmic fundamentals. Juxtapositions may help not only to explain but also to awaken fresh connections in an understanding of the logical way cosmos is developed.

Associative stacks are Natural Dialectic's backbone. The philosophy's three-mode motor is now up and running. Its polarities represent not only human but the cosmic spine; or, if you like, the system generates a muscular body of philosophy that accurately reflects the order of the cosmos. **It generates an abstract, metaphysical machine, tight-knit, well riveted by bolt and counter-bolt, the simplest working model of the universe.**

Chapter 6: Models and Stacks

Now what's to stop you building your own Primary and/ or secondary stacks? As you've seen, the columns run from *Sat* through *raj* and *tam* cosmic fundamentals. They run from potential (before) through action (now) to exhaustion (at the finish). In Chapter 4 we translated these into the language of mind (psychology), energy (physics) and mind-material body (biology). To recapitulate:

A simple link from <u>psychology</u> describes the three basic conditions of information as (sat) potential informant (concentrate of consciousness) prior to (raj) active mind (informant and informed) and (tam) passive, dormant or sub-conscious mind. Of course, information may be carried by non-conscious, material forms but it's essentially a mental entity.

A simple link from <u>physics</u> describes the three basic conditions of energy as (sat) potential prior to (raj) action and (tam) exhaustion.

And a simple link from <u>biology</u> describes the three basic conditions of life on earth as (sat) informative archetype, (raj) metabolism including a code carrier called DNA and (tam) product called body.

We now in a position to link models with stacks.

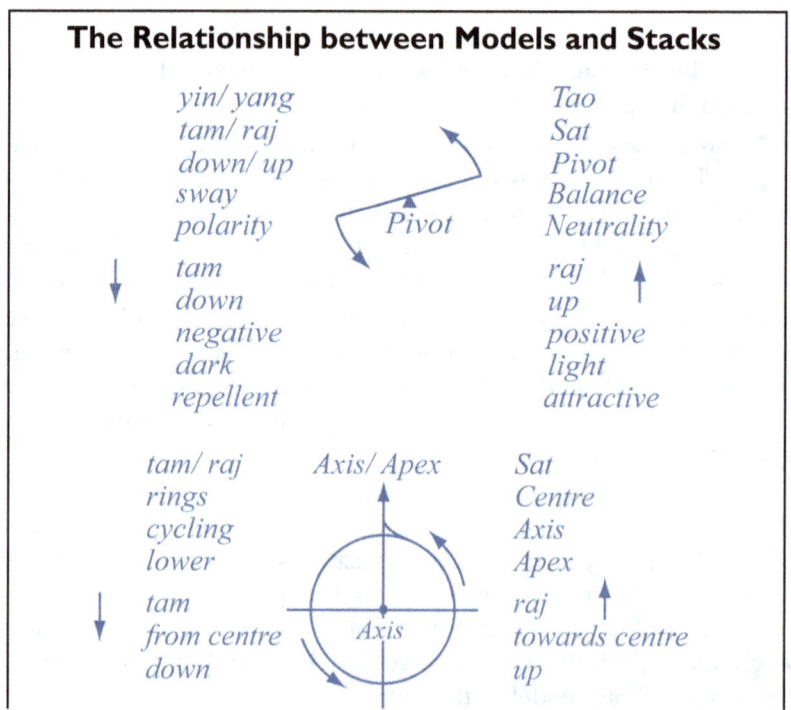

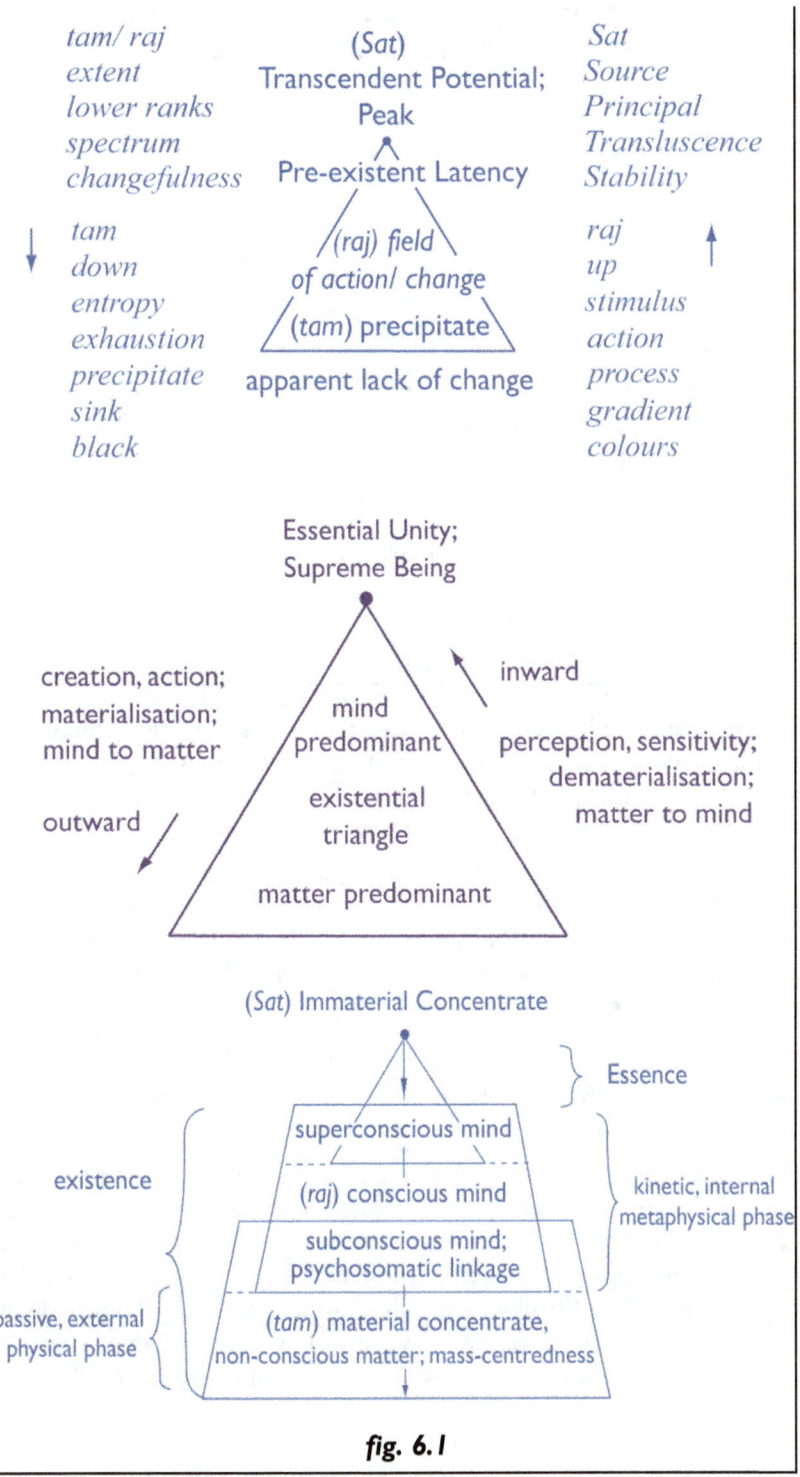

fig. 6.1

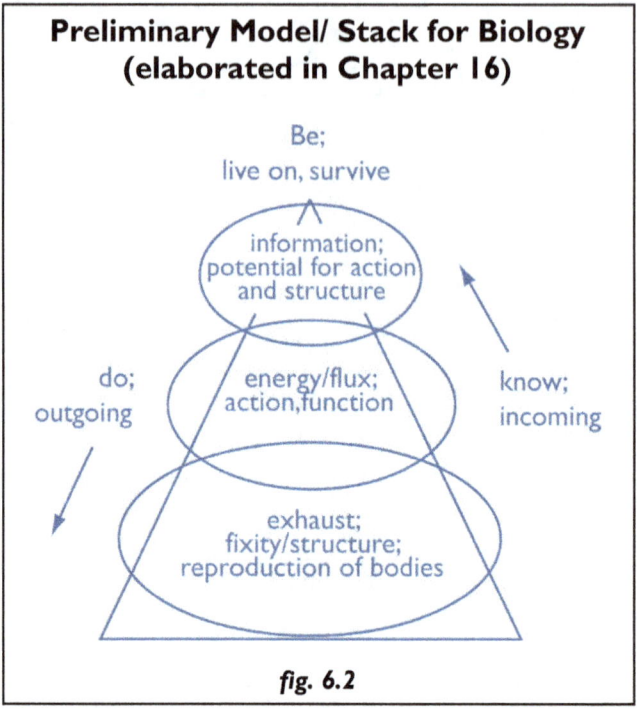

fig. 6.2

We have linked models with stacks and, recognising a hierarchy that runs between source and sink, are now in a position to begin to build a hierarchical picture of cosmos. The idea of cosmic hierarchy is currently unfashionable but we can now:

- note that, in the tri-logical view, pairs of opposites interact with one another under the influence of *three* cosmic fundamentals (up, down and equilibrating). Such triplex logic, embedded in all things, has always been intuitively grasped by seers, poets and playwrights.
- in Chapter 7 elaborate on the hierarchical perspective.
- in Chapter 8 relate Natural Dialectic's three main models, as above, to Essential and existential stacks.
- and begin to understand Origin. Firstly, think of (*sat*) source or potential. A stream springs running to exhaustion in the sea; and creation, as a 'field' called existence, must spring from a source as well. Existence is a river but this Pre-existent source, this Non-existent Equilibrium, we call Essence; that is, Being-without-predicate. The question is: what is the nature of this Void, this pre-conditional Nothingness? Is it material or immaterial?

Chapter 7: A Hierarchical Perspective

For Natural Dialectic materialism's single element, unconscious matter, yields a flat-earth kind of view. Let's take a tiered look at cosmos.

Hierarchical, Triplex Construction of the Cosmic Pyramid

The world, for a materialist, is basically interactions due to types of energy and force. This the canon, the creed, the belief.

If, however, you expand such atheist's primary axiom to allow an immaterial element then a holistic, three-dimensional view of cosmic hierarchy kicks straight back to life. Here's a stack resembling the Primary and Secondary stacks in Chapter 5.

projection	*Transcendence*
expression	*Latent Field of Potential*
flow	*Source*
existence	*Essence*
non-Essence	*Non-existence*
↓ *matter*	*mind* ↑
informed automaticity	*informant creativity*
non-conscious physic	*psychological motion*
sink	*flow*

And a triplex cosmic ziggurat.

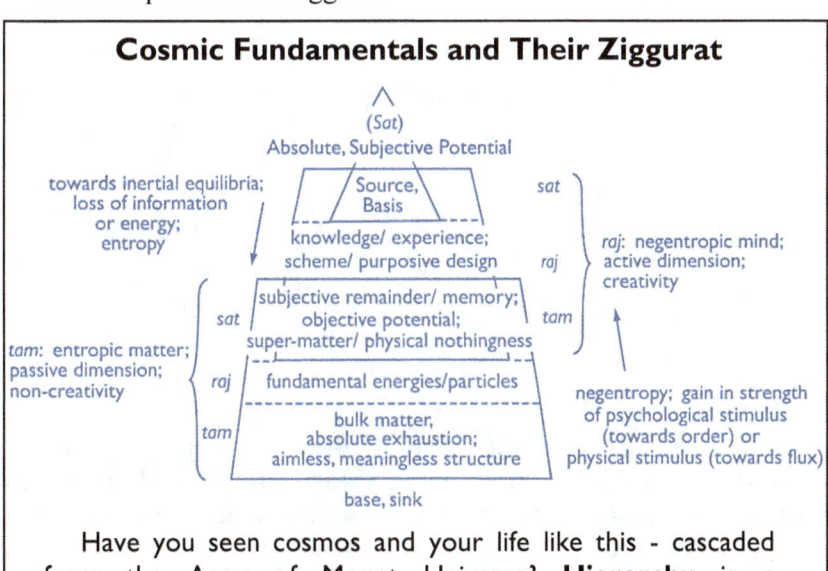

Have you seen cosmos and your life like this - cascaded from the Apex of Mount Universe? **Hierarchy** is a fundamental aspect of creation. *If you disagree with that, look no*

further than your own construction. It is (Chapter 15)[32] hierarchical.

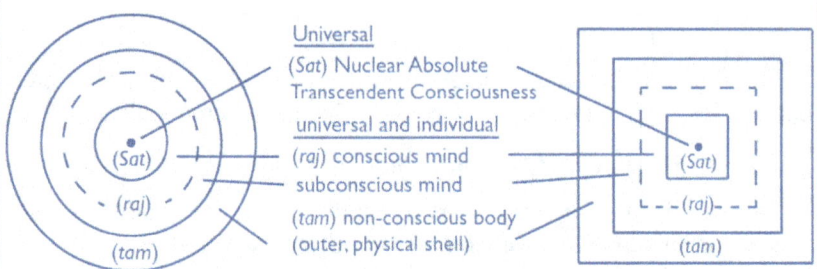

'Hierarchical' may also be construed as, like Russian dolls, 'nested'. In this case power is internalised towards the centre. Concentric circles spread from (*Sat*) Nuclear Cause. Where Apex becomes Centre a ziggurat, viewed from above, is represented as a squared nest; and concentric rings equally reflect tiered phases of, for example, an action of creation, cosmic structure or the development of a codified egg to adult form.

fig. 7.1

You can read each step of a ziggurat as nested in the next; from base to peak is, if you worked with concentric circles, from periphery to central source. In this case three broad hierarchical divisions can be pressed a little further into triplet sub-divisions of mind and matter.

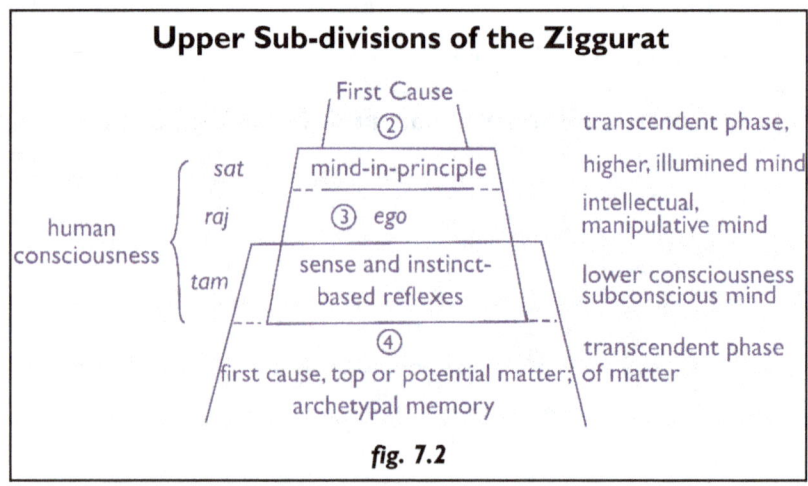

fig. 7.2

In *fig.* 7.1 the top *subjective* division, Potential, *is*. **Thence, downwardly, subjective sub-divisions at each level of mind** (in *fig.* 7.2) **represent *loss of active information* from the Latent Super-state;** that is, they represent lower, slower or, as with memory, fixed/ passive states

---

[32] also *SAS* Books 2 and 3; *AMA?* Chapter 24.

of mind. Such natural (↓) 'entropy of information' is consciously reversed by the (↑) negentropy of concentration. This rousing stimulus amounts to interest and, at root, intensity of focus called the fire of love. The highest aim of such a focus would, logically, be its Cosmic Source.

The psychology of consciousness and subconsciousness is discussed in Chapter 13.[33] The world of physics is appears in Chapter 14. In the sense that it 'underlies' mind our material universe might be termed an underworld.

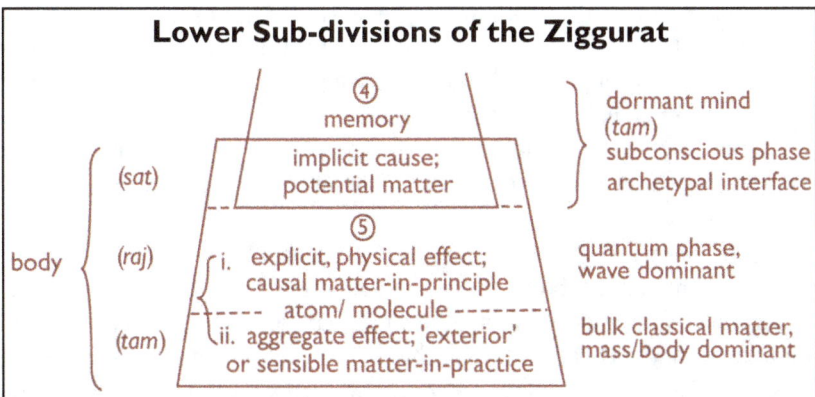

Potential matter is an implicit, metaphysical cause of physical effects. This transcendent sub-division of 'super-matter' and its psychosomatic interface are discussed in Chapters 7 – 15 under 'potential matter' and 'archetype'.

Through quantum physics could science touch with, at its implicit, causal or archetypal phase (phase 4), universal mind? Phase 5 involves all physical effects. Of these an explicit cause - cognate quantum particles and forces - gives rise to so-called 'condensed', 'externalised' or bulk matter. In other words, as science is aware, the material universe is an effect composed of interior, kinetic and exterior, inertial forms of energy. This couple are the subject of Chapters 7 - 12.

fig. 7.3

Objective sub-divisions of level 4 shown in *fig.* 7.3 represent internal, metaphysical, sub-conscious agency. **But external, bodily phenomena of level 5 are measured by a *loss of free energy*, by a downward 'thickening' of non-conscious materiality from the subtle state of plasma, light and quantum factors through gas and liquid down to gross solidity.** Such natural (↓) entropy is reversed by the (↑) negentropy of energetic input, that is, by variously concentrated

---

[33] also *SAS*: Chapters 5, 6 and 13-14, sub-consciousness in 15-17; and chapters 7-12 elaborate on the sub-divisions of non-conscious matter.

stimuli due to colliding momentum, fiery heat, explosion and so forth.

A physicist is well-placed to understand *fig.* 7.3. Although he has no way to measure immaterial potential he now knows (which previously he did not) that matter may involve an 'inner nature'. A quantum world (of atoms, particles and forces) underlies sensible phenomena. We can call this subtle foundation matter-in-principle and the gross appearance of gases, liquids and solids (our 'normal' world) matter-in-practice.

Physicists are also, therefore, well-placed to understand *fig.* 7.2. An individual's 'normal' thought is generally composed of a poorly controlled stream of instinct, sensation, emotion and calculation. But this superficial dynamic is supported by an 'inner nature' composed of deeper elements called mind-in-principle. Just as 'quantum' constitutes a 'higher' or 'more fundamental' level of matter so contemplative realisation of these natural principles constitutes a 'higher' or 'raised' level of consciousness.

These observations demonstrate that steps of the cosmic ziggurat involve elements of both science and philosophy. Material level is, according to the triplex order of cosmic fundamentals, connected with immaterial; and, also according to this order, subdivisions are arranged.

Such framework identifies a source and sink - a top super-state and base sub-state.[34] From its gradient we might, if the nature of First Cause is Informant, suppose that informative capacity degrades into the incapacity of oblivion. This would lead to an interpretation of non-conscious matter as a low-level projection of universal mind!

Finally, could you be construed as a microcosm that reflects the 'macro' way that nature works? Are you constructed hierarchically? Another way of looking at the **triplex cosmos** is indeed, as the next section illustrates, by reflection of yourself.

The Cosmological Axis

Who do you think you are and where?[35] What is the direction of your short-lived pass through time and space? A traveller wanting to arrive need first locate exactly where he is and where he thinks he's going. Every explorer needs to *orientate*. Only then can he plot with military precision the bearings of his destination and begin the chosen journey.

Check macrocosmic *fig.* 7.4 against microcosmic *fig.* 13.6.

To monopoly-materialism it appears that a human is an animal composed of chemicals. It is a child of chance, a minuscule creature uncradled in the vast, dark, cold and sometimes violent abyss of space. Such diminutive, deriving its morality in theory from mutant genes,

---

[34] this brace is explored in Chapter 9.
[35] also *SAS fig.*13.5: Cosmology of Your Psychological World.

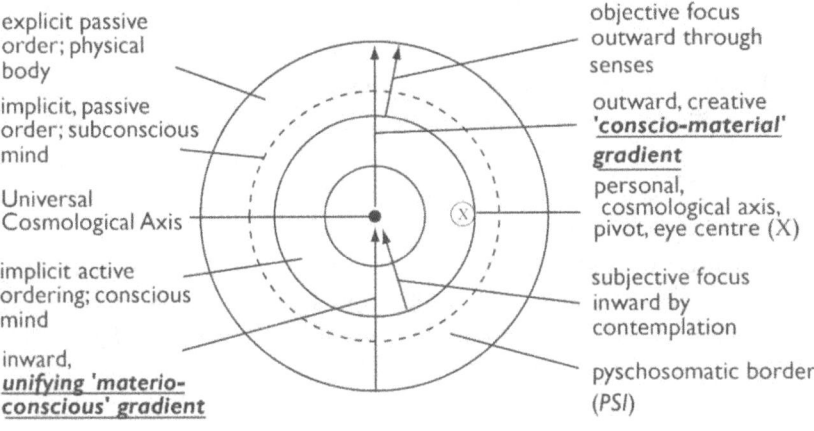

fig. 7.4

should entertain no thought of Central Origin. There exists no Upper Pole towards or from which humans turn.

Chemicals are part but are they whole truth? The holistic view is different. It includes a spectrum of subjective mind. Holism asks, *top-down*, the band of 'conscious visibility' in which I dwell. Where, between the poles (Transcendent Centre and subtendent body), do I stand? Three points on fig. 7.4's simple, nested model can act as a map to obtain cosmological bearings.[36]

First *is the psychosomatic (mind-matter) interface.*

X marks the **second** spot. Touch your forehead at a central point above the nose, just above and between the eyes. Just behind this, marked X on the diagram, is where you think. *This location, identified as the second point of interest, is called the eye-centre. It is not the central reality of the whole universe but it is certainly yours.*

From X, therefore, concentric rings of your environment radiate. Rings spread outward to a world of bodies - physical creation; but inwardly, tipped to the metaphysical direction, rings evolve (or perhaps we should say, *involve*) towards the Centre of their Concentration. In this respect X marks a mobile, personal pivot connecting inner and outer worlds. Thus the third eye can become the starting-point for inward, contemplative travel. *So, if you wish, X is transformed into the launch-pad towards a **third** point of interest - The Universal Cosmological Axis.* This Pole Star, Absolute Point of Reference or Pivotal Truth is within your own more outward X. Its Transcendent and Immobile Hub is *within* the ceaseless hive of mind; from it radiates a jubilant cosmology. Let us examine such Central Being.

---

[36] *SAS* Chapter 3: Cosmological Axis.

Chapter 8: Essence

Beyond existence Essence is Being.[37]

Such Essence is the first basic principal/ principle of Natural Dialectic. It is Potential without predicate. If creation can issue from it then, in this sense, it paradoxically contains everything. Its Potential is Omnipotent. Its Substance is Omnipresent.

A source is connected to yet also **independent** of its issue. The dependent issue would not happen without it. Thus Essence is, intrinsically, nothing existential. It is immaterial. It is not physical but metaphysical. Cosmic source is metaphysical.

tam/ raj	*Sat*
existence	*Essence*
becoming	*Being*
dependent	*Independent*
issue	*Source/ First Cause*
finite	*Infinite*
relative	*Absolute*
motion	*Equilibrium*
expression	*Pre-active Potential*
conditions	*Precondition*

The **Absolute Independence of Essence** (which is non-existent from an existential perspective) raises, in the relativity of ever-changing creation, Primary Paradox - of which Lao Tzu was so fond.[38] Its Being is in every being; its infinity is restricted into every time and place; it is everywhere yet nowhere, omnipresent and yet absent, neither psychological nor physical but projects the current that sustains both these dimensions.

If Absolute Essence transcends existential relativity yet is, at the same time, its Source then, as well as metaphysical starts, we are drawn by the Dialectic to metaphysical conclusions. Alpha and Omega are One. **Essential Unity** precedes all existential motion whether psychological or physical. Such Pre-active Independence is clearly immaterial.

natural	*Most Natural*
range of action	*Super-State*
dipoles	*Monopole*
polarity	*Neutrality*
with form	*Formless*
matter/ mind	*Soul*

---

[37] the word 'essence' derives from Latin *esse*, to be or *essentia*, essence.
[38] see Tao Te Ching: The Book of the Way.

If Essence isn't mind *or* matter it is, existentially, nothing. Paradoxically, the Source of everything is None of them. It is **Nothing**. Tzu's profundity of **Void** is Source of All. Thus Invisible Nothing is the centre-piece of Natural Dialectic; Supremacy of Void, empty of existential motion, is paradoxically full of possibilities; Potential's activation runs the cosmic show.

What more natural than Most Natural, the source of nature?

And who denies that nature's real? Or that its Top-Level Source must, therefore, be Most Real?

lesser being	*Being*
more or less real	*Reality*
relatively substantial	*Substantial*
local/ temporary	*Omnipresent*

How, though, does this Primary Dialectical stack make sense? Are all right-hand elements of such a Stack simply ways of saying the same thing?

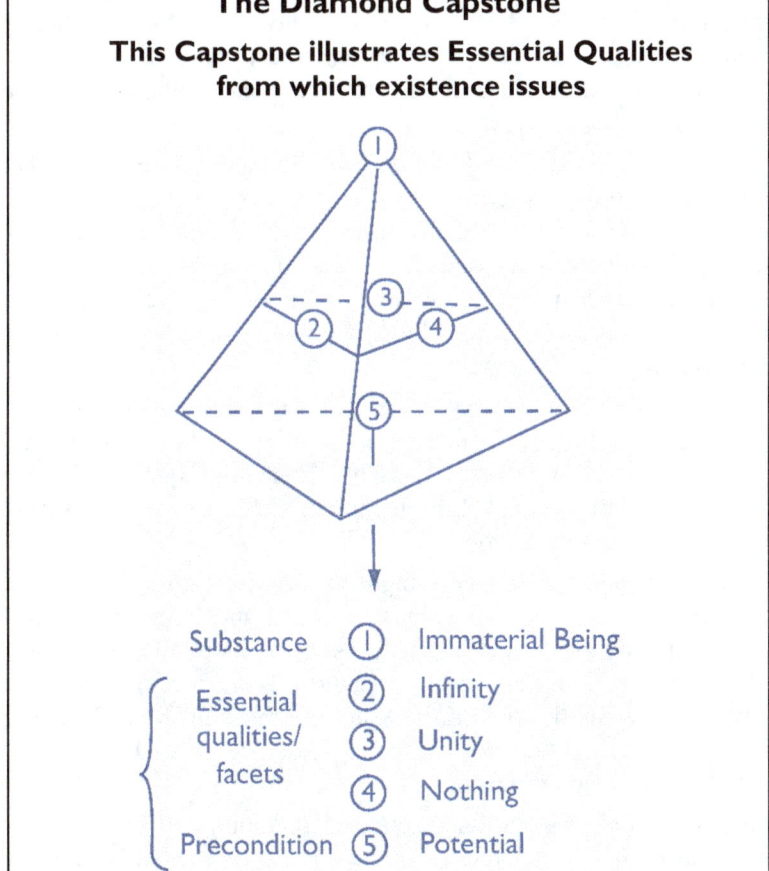

> The Apex of Mount Universe is likened to a Tetrahedral Diamond with three clear facets and, at the base of its transparency, a fourth to interface existence. The trinity involves, at the highest level, aspects of its Immaterial Substance. The root of immaterial information is Consciousness. Pre-active Consciousness, an Immobile Stability, is at the root of mental form. Such Concentrate, in whom all knowledge must potentially reside, is also known as Pure Life.
>
> Facets of the Diamond, Essence, include Infinity, Unity and Nothingness. These comprise Potential for First Cause from which creation is expressed; from this intrinsic precondition all the order of existence hierarchically falls. The nature of their reflection in the zones of conscious mind and non-conscious energy, that is, physics is investigated in Chapters 13 and 14 respectively.
>
> **fig. 8.1**

At this point, however, you may rightly object that for *bottom-up* materialism such a figment as Essence makes nonsense; and only the material fraction of existence - in fact, its whole - harbours sense. Imperceptible Supreme Being is an absurdity but what are 'lesser', perceptible beings such as local and temporary thoughts, objects and events? They are construed to have 'ordinary' being without superiority, inferiority or any sense of hierarchy.

For *top-down* holism, on the other hand, Essence and existence both make sense but only existence is sensible. Essence is paradoxical. It is beyond form and yet at once the heart of mind and body. It is the Unmoving, Uncreated Axis round and yet within which existence orbits; it is the Potential whence actuality appears. In your case everything you know appears in consciousness. How, though, did and does the universe appear; what is its fundamental field?

Duality within Unity. Existence is binary but contained within Singular Essence. Essence is nothing existential and yet is the cause of existence. It is 'beyond' yet creates the existential system. From an absolute viewpoint (called Enlightenment) all is, within Essential Projection, quintessentially one.

However, a second objection might be that creation is littered with various essences, potentials, balances, voids, units and neutralities. There continually occur innumerable cases of inaction, formlessness and being *within* the relativity of existence. Therefore, on what basis can an Absolute Reality or Infinite Equilibrium be proposed? In short, when essential characteristics commonly exist, on what basis can the Essential side of Primary Dialectic be sustained?

Yes, the *Sat* or Essential fundamental *is* commonly reflected as part of creation *but only in local, temporary forms*. Every existential object or event has boundaries; these bestow psychological (in the subjective case of mind) or physical (in the objective case of bodies) condition. Thus we call such realities 'lesser', 'seeming' or 'apparent'. They are finite in some degree. They (including balances, potentials and so on) are relatively restricted, that is, localised in time and space.

On this basis the ceaseless flux of existence is construed as an appearance, a projection or an output of Essential Source.

So 'lesser' units always involve flux and boundary. 'Lesser beings' constitute all existential forms of energy, matter and mind bound in time and space. They are 'phenomena', a Greek word meaning 'appearances'.

Time and space are interesting lesser infinities. While creation is composed of objects and events, these peculiar 'containers' or 'extents' appear boundless and consist of physical nothingness. They are immaterial but intangibly support all forms. They are essential for existence but are they *per se* the source of worlds? In other words, was there time and/ or space *before* creation? If not, they are not Essential. What is the nature of their lesser essence? [39]

More or less implies a scale. Before inspecting 'in-between' what is at the top and base of cosmic scale?

Source and Sink

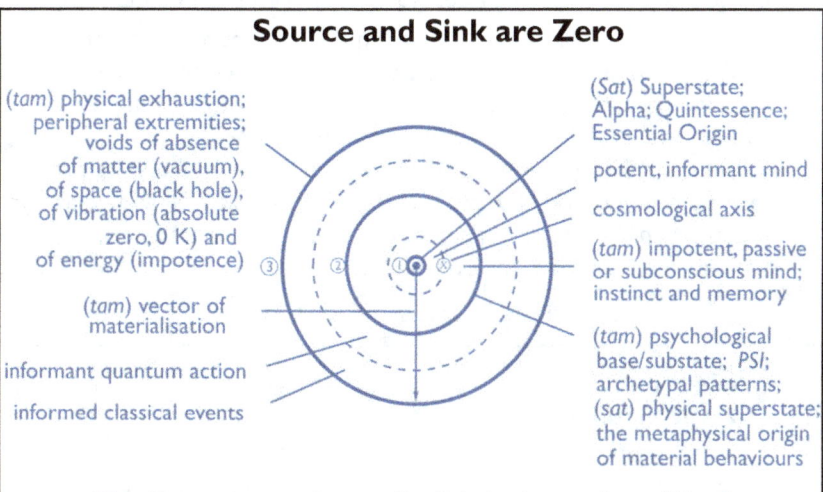

The illustration rephrases fig. 7.4. It shows three '0's, three kinds of zero.
  (1) Essential Source of Informative Principals
  (2) psychological sink (archetypal memory)
       metaphysical source of specified forms of physical energy
  (3) physical (and existential) sink

---

[39] see Chapter 14; also *SAS* Chapters 9 and 10 for a fuller appreciation of this brace of omnipresent mysteries.

In one sense, therefore, you can understand that nothing is a latency; *source* is potential; it is 'nothing happening yet' before expression of the possibilities. In the other sense there's nothing left; exhaustion is the end of possibility; zero can be a sink as well.

Alpha Source (1) is Central and has two 'zero anti-poles'. One is subjective/ psychological, the other objective/ physical - in that order.

The subjective anti-pole is 'zero psychological' (2). It is a sub-state sink of unawareness, sleep and fixity of thought; it includes the passive and subconscious mode of mind called memory.

Psychological sub-state is, paradoxically, at once material super-state. In this Janus-like sense it is called potential matter (*figs.* 7.2 and 7.3). From such metaphysic physic orderly derives. Didn't starry cosmos spring from nothing-physical; don't natural phenomena emerge from zero? Fixity of *mind*, archetypal memory, serves as the source of information. Material government derives from this 'great instinct' of the universe. So memory's zero-physicality is metaphysically non-zero; its substantial, archetypal templates regulate the character of every automatic, physical event. Potential matter is not physical but its higher, immaterial priority controls the way materials non-consciously behave. Starry cosmos sprang with this in mind.

A hundred billion galaxies still make subjective zero. Non-conscious matter (3) is the anti-pole of Immaterial Essence (1), mind and material archetypes (2); and the extreme of matter's anti-pole is its objective zero. This sink, 'zero physical' (3), marks total drainage at creation's furthest edge, exhaustion at its dark periphery. Its rim marks an extremity of Alpha's absence. It is matter's end-point, negative annihilation, empty space; its subtendence thereby marks a natural posterior and rump of universe.

fig. 8.2 (see also fig. 7.4)

tam/ raj	*Sat*
relative nothings/ apparent things	*Nothing*
projection	*Potential*
lesser beings	*Supreme Being*
issue	*Source*
derivations/ dependencies	*Super-State Void*
↓ *tam*	*raj* ↑
zero/ absence/ space	*entities*
sub-state impotence	*events*
dead stillness	*change*
finish/ sink	*process*

The upper and lower extremities of cosmos are its Source and sink. These, transcendent and subtendent, will be discussed in Chapter 8.

In between extremities 'more or less' of either implies the presence of a scale of relativity.[40] A conscio-material gradient of Mount Universe, whose ultimate Criterion is Essence, is illustrated by *fig. 7.1*; and *figs.* 7.2 and 7.3 show how each major subdivision hosts its source and sink as well. These, lesser essences, would scale away from Top to bottom. What, we ask, is the Essential Nature that creation gradually leaves? The pure non-consciousness of matter? Or Pure Informative Potential viz. Consciousness?

Specifically for mind (nothing physical) we read:

tam/ raj	*Sat*
states of mind	*Informant Logos*
oblivion	*awareness*
unconsciousness	*perception/ cogitation*

and for energetic matter (nothing psychological):

material states	*Potential Matter*
vibrant forms	*Informative Archetype*
absence	*presence*
non-locality	*locality*
general space-time	*thing*

The link from physics shows three phases every form of object or event traverses. It runs, more or less locally and temporarily, from (*sat*) potential through (*raj*) action to (*tam*) completion. But, even though they may embody the same characteristic of immobile nothingness, Potential and impotence are diametrical opposites.[41] They mark precursor and the end of any process. Source and sink. Both states might appear restful but, as the next section illustrates, each kind of void is the *inverse* of ts its antipode.[42]

Voids

Cosmic source and sink are super-state and sub-state voids.[43] Voids are nothingness but the nature of these two extremes is very different. Their character is opposite in every way.

Take zero's diametrically opposing forms. In a *plus way* nothing comes before. It holds the future. Void precedes. It represents potential, poise or capability *before* the action starts. Such *positive* zero describes

---

[40] Chapter 16; also *SAS* Chapter 4: Truth, Appearance and Reality.
[41] *SAS* Chapter 9: Dialectical Vacuum.
[42] also *fig.* 10.3 Inversion (from archetypal super-state to exhausted sub-state).
[43] also *SAS* Chapter 9: Nothing.

the state of pre-condition that anticipates an actual behaviour and without which the behaviour, pattern or form of event could not occur. A tiger is poised to pounce. Such poised readiness is the source of possibilities prior to the realisation of any particular one. **Again, potential energy precedes kinetic energy; it precedes any particular creations of that energy and, of course, its exhausted effects.**

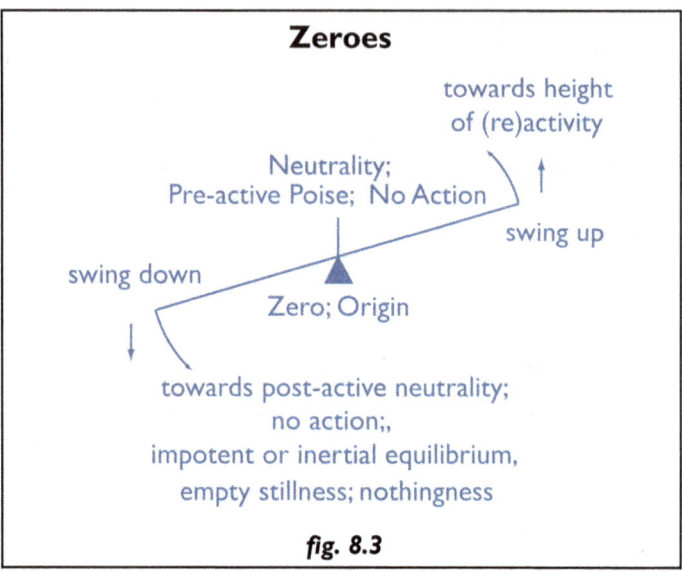

fig. 8.3

In a *minus* way information and/ or energy has been locked or lost. Absence also describes a state (or, rather, statelessness) that occurs when everything has been subtracted; or when nothing has been added to nothing in the first place. Zero.[44] *Negative* zero is a bereft sub-state of exhaustion, abstraction or impotent void. Nothing left comes after. The bottom of the pile's a sink. This is the neutralisation, the annihilation, the zero of death.

In brief, you can think of nothing in two crucially different ways. **Potential** and *exhaustion*. **You can find potential and exhaustion, pre-active and post-active absences of change.** There's everything to play for prior to when the game begins; and nothing left when it's all over. *Within incessant cosmic motion only lesser potentials and impotence occur. Such relative events are seemingly inactive but also local, specific and impermanent.* Poised or spent, specific inactions are therefore termed lesser absolutes, relative neutralities or special cases, zeroes, in a world of motion. Such lesser stabilities appear within the flux of change. But while (like rules of natural behaviours or empty space in which to play creation's energetic game) they reflect permanence, could there exist Substantial Stability? In a hierarchically structured universe

---

[44] see Glossary: non-existence; and *SAS* Chapter 9 and Index: zero/ nothingness.

could there be Supreme Potential sufficient in its generality to 'fire' the certain set of possibilities that we call cosmos?

finite	Infinite
1/ on/ action	*0/ Off/ Peace*
object/ event	*Nothing*
polarity	*Neutrality*
duality	*Unity*
↓ *post-active effect*	*causal stimulus* ↑
no-change/ apparent fixity	*changes*
switched off	*switched on*
unit/ object	*flux/ event*
unreactive/ neutral	*charge-based reaction*
peace/ zero action	*action*

Let's rephrase. If you believe the universe was always here, there is no need for any start. However, modern science no more subscribes to eternal matter than holism to a never-starting, never-ending cosmos. If, therefore, existence had a start then nothing, at least nothing existential, must have come before.

If, in accordance with materialistic culture, you believe that only physical phenomena once 'started up', what kind of non-physicality projected physicality? *What kind of 'immaterial nothingness' preceded 'anything'?* If your starter is a pyrotechnic of dense radiation and expanding moment-volumes of space-time what triggered this appearance?

On the one hand, nothing flows through nothing. Absence by itself cannot create a presence and, the fact is, pure impotence (like space or time) can never make a thing. Before things (and thoughts are things as well) there were none at all. Perceptions, time and thoughts are the ingredients of mind. And space, time and things are the ingredients of physicality but if these came from nothing what's the nature of preceding absence? Can you really claim that absolutely nothing nowhere for no reason rippled; a spontaneous ripple swelled *ex nihilo* and from this inflation everything exploded - not least, given time galore, you, me and all else that lives? Full marks for a magnificent mythology!

On the other hand, if you believe existence (mind and matter) stems from a projection, what holistic kind of 'nothingness' preceded it? What is the nature of Transcendent Void whence cosmos was projected?

Of course, the *top-down* notion of such psychological/ metaphysical projection grates on materialistic nerves. *In terms of Natural Dialectic Nothing (truly Essence prior to all existence) is Potential.* Thus potential-prior-to-action is of prime importance. Either way, for holist or for humanist, the gift of everything depends upon the Nature of Prior Nothingness!

hierarchy	*Top*
impurities/ relative purities	*Purity*
appearance	*Truth*
conscio-material gradient	*Pure Immateriality*
objective-subjective mix	*Pure Subjectivity*
scale of existence	*Essence*

Let's, for a second time, rephrase. A principal comes first in line; a principle packs information, source of order or a guiding force. What has form exists; it has a start. If neither mental nor material form exists there's nothing - their non-being. Void is neither any thing nor form of thought at all. Yet such prior Nothingness, from which they must have formed, is The Essential Being. Call it metaphysical necessity. Grasp, therefore, the first and final paradox - non-being is The Being! **Thus Fundamental Nothing is the Principal and the Essential Principle of Natural Dialectic.**

Nor, again, should use of the same word to signify different conditions confuse us.

tam/ raj	*Sat*
action	*Inaction*
restlessness	*Rest*
swing	*Balance*
↓ *tam*	*raj* ↑
inaction	*action*
RIP/ death	*liveliness*

For example, the words peace, rest and inaction are used top right and bottom in stacks but in each case they mean opposite things. They constitute antipodal opposites. Their characters are positive and negative. In short, there is extreme antithesis between positive and negative zeros, that is, between the 'Super-state' of fully unrealised potential and its wholly exhausted expression called a 'sub-state'; also between potential and exhausted equilibria and immobilities. The 'flip' from one extreme of a character to its opposite is called a Dialectical Inversion.[45]

Top Positive and base negative represent extremities of scale. Essential Purity is cosmic source but what is sink? In the next chapter we'll consider states of extremity wherein one state above another is called transcendent and below is called subtendent. Both relative and absolute species of these states exist. How can such absolution be?

---

[45] see *fig.* 10.3 and Chapter 10: Polarity.

Chapter 9: Extremities

The world includes superlatives - most high, very low and so forth; and its scales sport extremities called tops and bottoms. A scale is effectively a hierarchy that reflects, either in continuous or discrete mode, an informative or energetic level. It may, like the electro-magnetic spectrum, include designated sub-sections such as UV, IR or visible light. *A level above transcends and one below subtends.* So UV transcends and IR subtends visible light; a gas transcends and solid subtends liquid. These are lower or minor transcendencies or subtendencies. Cosmos, in both its physical and psychological aspects, involves myriad different kinds of scale.

Is there, however, a Transcendence or an Absolute Subtendency? If, on any gradient, you can rise or fall how far, in each direction, can you go? What are the practical limits of the (↑) upward and the (↓) downward vectors of the Cosmic Scale? What *are* the peak and base of Mount Universe? These would constitute the boundaries of mind and matter. They would edge the universe. To find them we start by tracing subtendence to its base extremities.

Subtendence

tam/ raj	*Sat*
below	*Transcendence*
range of action	*Super-State*
expression	*Pre-motive Potential*
current	*Source*
↓ *tam*	*raj* ↑
descent	*ascent*
subtending	*transcending*
deactivation/ fixation	*rarefication*
subtendent base/ sink	*stimulus*

The (*tam*) cosmic tendency is down from peak and out from centre. To subtend is to extend beneath or to occupy an adjacent, lower position. Natural Dialectic employs the term to indicate a lower level of creation. On a cosmic scale, to descend is to reduce, restrict and deaden. It is to lose energy and/ or information. Thus, to describe Natural Dialectic's negative extremes we can reasonably coin the word 'subtendence'. At most extreme a sub-state pole diametrically opposes a transcendent super-state. Base is the opposite of peak, sink the opposite of source. Is either state an absolute? What is the nature of our various universal sinks?

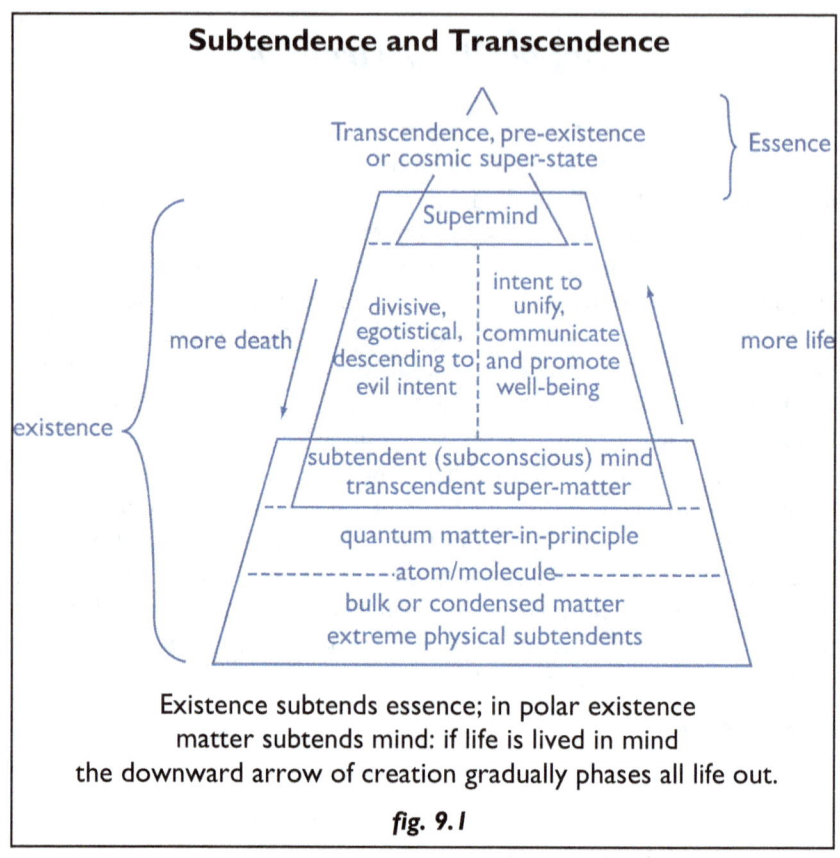

fig. 9.1

It is true, *top-down*, that products subtend their producers. A producer stimulates and stirs things up. From this perspective information precedes its material consequence, metaphysical mind plans physical arrangements and super-natural precedes natural order. *Mind first, body after; body's an appendage of its mind.* Does it, that physical depends on metaphysical, invert your 'normal' sense of things?

In the psychological zone sleep, we say, subtends waking; *subconscious subtends conscious* and the latter subtends superconsciousness. Base psychology involves oblivion.

Although we may be unaware of it, the sub-conscious, dormant level of mind still holds information, informs and is informed. In humans it is the repository of memory, instinct and the psychosomatic (*PSI*) channels which interface, immediately and continually, with the nervous system and thereby molecular structures, cells and organs of the body. *Stored, reflex or automated information is passive; passive information subtends active.* A singer records a song. The CD carries rigidified information. What is it but a plastic form of memory? All matter, which is totally non-conscious, repetitious and predictable in its behaviours, is informed by the internal or immanent exigency of 'laws of nature'. What is this 'song', whose record plays across the fabric of the known universe? Material bodies are governed by and thus subtend its vibrant rules.

Let's take your own body first. When you waggle your little finger, is it possible that the line of operational command runs from conscious mind through sub-conscious templates (memories, instincts etc.) over a psychosomatic border? And that incoming or outgoing information would be first translated to, or last translated from, the physical side by the product of 'excited' electrons, that is, electromagnetism? Where *matter subtends mind*, this 'radiant' phase would be subtended by the biochemical; and both levels occur within bulk, biological structures, that is, cells, nervous system, muscles and a coherent whole body. This places your finger-waggle at the base of an informative/ energetic hierarchy. Your sensible, physical form (perhaps including the origin of its shape) equally resides at the bottom of such a hierarchy. It would be the final outcome of a two-pronged, coded plan. First prong, generic, springs from an informative program called archetype;[46] the second, locally specific, is genomic. From signal biochemistry is body built; phenotype subtends its molecules and genotype. Thus bio-form's a frozen yet dynamic program; or, prosaically, a 'functional structure'. Very complex, yes; automated, yes; it's an incredible machine.

All bulk bodies subtend quantum agents; these 'organise' their chemistry. Whence, from a *top-down* angle, did cooperative arrangements of these subtle agents rise? What is the template they in turn subtend? If one does exist how does connection to it work? Could, closest to whatever transcends physicality, pure energy (say, mass-less radiance of light or binding nuclear force) subtend a signal interface? In consequence bulk, solid rock 'descends' from gas (which, if the universe was once much hotter than it is today, it did) because the gas itself subtended originally-projected, pre-atomic energy: the character of force-with-particles is ordered under archetype. From Natural Dialectic's point of view (*figs.* 7.3 and 9.1) quantum physics interfaces archetypal memory; *non-conscious matter naturally subtends a psychological potential.*

If (↑) vaporised concretions rise as gas the (↓) opposite applies. Hence the question:

'What's the end precipitate of gas, the final stage of matter past which it is impossible to drop? Of what consists the world's subtendent end?'

At the outer rim of physical projection, energy is frozen; any solid represents creation's edge; it represents, full stop, world's end. And this lifeless, oblivious underworld we call the zone of science expresses two forms of extreme (*tam* ↓) subtendence, one energetic and the other

---

[46] see Chapter 15: Conceptual Biology and Glossary; also *SAS* Chapters 16 and 19: Conceptual Biology.

massive. An object, super-cooled, approaches (but can never quite achieve) a total 'loss of levity', an immobility called 0°K.[47] The second, massive kind of cosmic death involves approaching total 'gain of gravity'. Extreme concentrations of matter include neutron stars, hypothetical 'quark stars' and a subtendence throttling matter past the nth degree called a black hole. Is this subtendence (ring-fenced by a 'horizon' that stymies any probe) infinitely dense or not? At least, utter negativity makes sense but are its boundaries permanent and real? Some are sure but, in the case that starts and ends are veiled in mystery, men interpret while nobody really knows. How would you, a scientific super-hero, break beyond the bounds of space and time and things?

To summarise: diametrically opposed to Essence exist some extremely negative polarisations. Such subtendencies are due, in the informative case, to severe loss or paralysis of consciousness and, in the energetic case, to extreme loss or confinement of energy. Yet such exhausted sinks are localised. They come and go. Change-prone relativities do not contain The Absolute. There can be neither Subtendent Purity nor Perfect Negative[48] where, in cyclical existence, all is relative and nothing, even any black hole, lasts forever.

Transcendence

	existence	*Essence*
	lesser	*Supreme*
	relativity	*Absolution*
	lower	*Highest*
	subtendent creation	*Transcendence*
↓	*downward*	*upward* ↑
	lowest/ lower	*higher*
	outer	*inner*
	inferior	*superior*
	contraction	*action/ release*
	fixity/ sink	*flux/ stream*

---

[47] 0°K is a strange, paradoxical cosmic boundary. Such absolution is, according to the Third Law of Thermodynamics, impossible to obtain but its proximal range involves properties of super-conductivity, super-fluidity and apparently infinite thermal conductivity! Such a freeze lacks any energy; its empty edge contrasts the source of physical creation; its sink of nothingness is fallen polar opposite of big-bang's sudden super-force and absolutely-nothing's miracle of wholly energetic ultra-heat (perhaps 10^{28} °K). See especially *AMA?*: Chapters 9, 11 and 12.

[48] *SAS* Chapter 26: Nature's Negativity and The Nature of Evil.

The *(raj)* cosmic tendency is up towards peak and inwards from periphery to centre. It involves concentration, that is, gain of energy or information. To transcend is to extend above or occupy an adjacent, higher position. Of course, as with a gas transcending its liquid or one person being brighter or prettier than another, the word means no more than increase on some energetic or other scale; they are lower or minor transcendencies. But materialism's lexicon eschews the word 'Transcendence'. If everything's material then what but nothingness could possibly transcend it? Then ask, again, what might be the nature of Transcendent Void? Did Chapter 8 not offer clues?

	matter/ mind	*Transcendence*
	outworking	*Ideal*
	comparative	*Superlative*
	relativities	*Full Life*
	lesser perfection	*Perfection*
	lesser positivities	*Summum Bonum*
↓	*dark*	*light* ↑
	decreasingly positive	*increasingly positive*
	sinister	*dexter*
	closed	*open*
	diversification	*unification*
	isolation	*togetherness*
	hate	*love*
	lie	*truth*
	confinement	*freedom*
	matter/ body	*mind*
	more death	*more life*
	sublative lifelessness	*life-in-relativity*

From *figs.* 6.1, 8.1 and, in the next Chapter, *figs.* 9.1-3 we can understand that a Dialectical illustration of the universe includes two hierarchically arranged main components, psychological information above physical energy.

In ascent, therefore, let's first treat physical transcendence.

Dialectical (↑) ascent translates, in dissolution, to release, greater freedom, individual forms lost in assimilation with a larger whole and loss of mass. Aren't these are, at physical extreme, attributes that reach towards the character of mass-less light?

Transparent, insubstantial - who can shed some light? Do you think a photon you can't squeeze between your fingers isn't powerful? Light transcends mass and, almost immaterial, flies with absolute velocity.

Pure energy, whose speed cannot be overtaken, represents an 'upper boundary' yet, concentrated, its ethereality can cut through steel. 'Grounded' it transforms to heat; its 'nothingness' drives life on earth. And illumination is communicative, it is information-bearing - you can signal using it.

'Potential matter' means 'where matter comes from'. Could brilliance, with its transcendent tendencies, interface with potential matter, that is, with the source of physics? Could radiant and contractive forces link physicality to what it's not, that is, to metaphysic? **Electro-magnetism linked with electronic charge is, for Natural Dialectic, physic *nearest* unto metaphysic - that is, to transcendent or potential matter.** [49]

Let's rephrase. Quantum action forms the basis of our universe. It forms the letters, punctuation and dynamic grammar from which grosser objects all appear. Or, where the analogy is music theory, it resembles vibratory notes on nature's scale. In this way could not quantum particles and basic forces be the medium for influence of potential matter, super-matter or informative programming called an archetype?[50] And, if the purest form of energy is light, could not purity (of energy or, psychologically, information) illuminate the borders of transcendence. Illumination would inform the interface from material to mental and, higher up, from mental to Supreme.

From quantum subtlety we pass to metaphysical phenomena. Mind's scale is, like matter's, subdivided. On the analogy of spectrum an ultra-conscious band of mind transcends our perceived state of consciousness which, in turn, transcends infra-consciousness. In other words, analogous with steps of matter, (*sat*) super-conscious transcends (*raj*) conscious which in turn transcends (*tam*) sub-conscious phase; and each mental phase exhibits its own properties. For example, nothing means anything to inertial sleep; indeed, sub-consciousness seems nothing either when I sleep or wake. Is such mundane transcendence, waking over sleep, as far as waking goes? Are you, locked to a body-shell, fully woken up? You feel so, as does fish or bird or bee, but are the brain-trammelled states of your own mind the only possibilities? Could further, voluntary awakening transport you nearer to Transcendence; might it lift you up the grades, across constraints towards excellence and sweep you ultra-consciously towards The One?

First out, last in. If the first expression of Perfection was First

---

[49] *figs.* 7.3, 9.1, 11.2 and 11.3.

[50] see Glossary; also *SAS* Chapter 5: Information's Infrastructure, Code and *fig.* 5.1; Chapter 11: Matter's Holy Ghost; also Index: archetype, alphabet, code, cosmo-logic, cosmo-logical language, information, hierarchy and potential.

Cause,[51] the last stage of return will know The First. It will realise (*fig. 7.2, phase 2*) The Cosmic Archetype. This is the first, primary current of creativity called *Logos*, Holy Name, Christian *Word*, Koranic *Kalam-i-Illahi* or Sufic *Kun*. As the first vibration from which creation emanates it is Sikh *Shabda* and Hindu *Paranada* whence *Om* and other *Nadas* (sounds) descend. As befits our 'information age' call *Logos* The Informant. Hence gradual (↓) materialisation of a cosmos orderly ensues.

Ascent is the anti-parallel direction. Informative and energetic foci always stimulate. They (↑) dematerialise to different degrees. *Raj* moves, experientially, towards Lightness and, informatively, towards Illumination. The ultimate dissolution, the most important transcendence is logically from existence to Essence. This step, beyond even the 'Living Sound and Light' of First Cause, is variously called Communion, Release or, in Hebrew from the Jewish faith, *Ain Sof*.

The idea of scale has led directly to a hierarchical perspective of creation. It is to a more detailed consideration of this that we now turn.

---

[51] see Glossary: first causes; also *SAS* Chapters 5: (*Sat*) Potential or Transcendent Information and Top Teleology; and 14: First State of Consciousness, The Psychology of Transcendence.

Chapter 10: Existence

Any engineer will tell you there is no avoiding first principles. However abstract or impractical they may appear, they constitute the basis for whatever system follows. If Natural Dialectic's structural design is symmetrical, strong and works then details should fall into place; if it accurately reflects nature, including mind, it should be functional; and, although its logic is not typically mathematical, it should be self-consistent. It should 'work'. Such 'functional logic' should provoke patterns of thought, questions and, within its framework, solutions.

Existence is the second basic principal/ principle of Natural Dialectic. It is, according to the Dialectic's Primary Axiom, a compound of mind and matter. It is, therefore, the umbrella-word for all psychological and physical constructions. It includes every formulation and formation, that is, everything. Creation is, paradoxically, both within and without Essence. **Of the Essential Pair it is the 'other', non-essential and dependent pole.**

As a dependency it has been caused. Causality is a first principle.

Causality

existential duality	Essential Unity
lesser sorts of being	Supreme Being
conditions	Precondition
issue	Source/ First Cause
↓ effect	cause ↑
physical bodies/ events	mind
complete materiality	shades
result/ product	excitation
counteraction	action

Aristotle believed in a First Mover itself unmoved by any cause. St. Augustine observed that no 'efficient cause' can cause and thus precede itself. Thus causal order can't be infinite; there needs to be an uncaused primal cause. Existence is composed of caused, finite events. Whatever begins to exist, asserted the *sufi* Algazel, has a cause; and 'something which begins has a sufficient cause' is also the modern principle of causality. This principle is constantly verified and never falsified. The physical universe began to exist and therefore has a cause. What is caused is not eternal. It is finite. Its effect becomes a further cause. Thus all existence is a changeful network made of causes and effects; creation is an action and reaction zone.

Where nothing is an absence, nothing comes from nothing. Did you think space was nothing? Wrong. Vacuum did not come before its cosmos; nor, since particles arise from fluctuations in it, is it nothing. Why, therefore, should cosmos as a whole derive from nothing? If it didn't, did the universe create itself? Then it existed prior to itself. Cause caused itself. Such, since no thing can create itself, is the kind of incoherent, 'boot-strap' logic some these days display. Perhaps, you claim, there was no cause of physicality! But, if there were, such cause could not be physical. It must transcend the physical. Its physical non-being must be immaterial; its being must be metaphysical. This Uncaused Being is not absent. It is self-sufficient, potent and with presence; its causal level of reality is 'higher' than non-conscious, physical phenomena; and its timeless time is Now.

In summary, it's as simple as it's crystal clear. **What starts to exist is always caused. Material cosmos, known as nature to the natural sciences, started to exist and therefore has a Natural Cause**. This cause, preceding physicality, is physical non-being. It is nothing physical but causes cosmological effects. What came (or comes) before the latter's matter, space and time must itself be time-less, space-less, immaterial. **Preceding its own secondary causes and effects the primary, first cause must be physically uncaused; it is uncreated in a naturalistic sense, thus super-natural. Metaphysical. In this respect the term for cosmic origin preferred by Natural Dialectic to 'big bang' is '*transcendent projection*'**. Since there's no scientific explanation you might reasonably term such origin a miracle. Shot from its archer's super-natural bow this cosmic missile undergoes, according to time's arrow, changeful yet intractable decay. Why should what precedes the arrow ever suffer *physical* demise? Better try and understand the nature of an uncreated immanence from whose causation galaxies and men have sprung.

Rephrased, motion is potential that's expressed; cosmic action is Uncaused Potential worked.[52] The primal motion of Essence is First Cause; such First Expression is the Start of starts and purest form creation issues in. All existence, psychological and physical, is the Start's effect. Although, within the whirligig of space and time, we choose to call some actions 'causes' and others their 'effects' in fact they're all, traced back to origin, effects of Cause. **First Cause is linked with everything that follows its beginning; and it forms the link between the pair - Essence Absolute and relative existence - that comprise the basic, prime polarity of Natural Dialectic.**

Since the energetic universe was once initialised creation's been

---

[52] check *fig*. 3.3.

replaced by conservation (of energy) and transformation.[53] Transformations are what science tracks; and Newton's Third Law of Motion (that every action has an equal and opposite reaction) captures the intrinsic order of equilibration underlying changes in this natural traffic. The law grants cosmos fundamental balance and descriptive science its equations. In the orient it's translated, where '*karma*' means 'action', into the '*karmic*' law of chain reaction.

Why, however, should a balance sheet of the world's economy not extend to the psychological dimension? Why should precision in the automatic world of matter not be mirrored in the mind - karma physical *and* psychological? As with physic so with metaphysic; every desire provokes an equal and opposite reaction; causal thought creates a boomerang effect.[54]

You understand such psychological equilibration well. Scales and sword. Natural *karma* is reflected in a golden statue. Objective and impartial Lady Justice tops Old Bailey; and she represents, in principle, all other human courts.

Cause informs effect. At this point let us briefly introduce a mystery that through the book we shall increasingly explore. **No doubt, energetic causes push effects; they bump you from behind; their arrow, physic's arrow, runs from past to present.** And things suffer (though you'd hardly think it as regards a proton or electron) from increasing weariness called entropy. They run out of steam. What, though, about a cause that is conceptually implanted? What about *informative causation*? **This is goal-oriented; and goals are in the future pulling you their way. They pull you from the future; they lift forward. They are metaphysical *attractors*, guides that govern your behaviour as they lead you through the world.** Not material but immaterial, such leadership is not by force of gravity, electric charge or magnetism; information is a metaphysical not physical affair; it's psychological and so its space and time are not the same as physic's. *Information's entropy is negative.* **Mind is negentropic and thus metaphysic's arrow flies, from future back to present, anti-parallel to physic's.** Thus you are guided, present to the future, by plans realising goals.

Purpose is what pulls you personally forward; and, even if they're fixed, we'll see how creation's vibrant generalities, its behavioural attractors, pull things forward the way they have to go. **Push and pull cooperate. Energetic and informative causation, running anti-parallel, are the way the world proceeds.**

In future we shall not much use the word 'attractor'. **The Dialectic's**

---

[53] *SAS* Chapter 12: Labours of an Empty Womb.
[54] *SAS karma* theory; see also Chapter 18: Anathema, Chapter 26: Social and Individual Parts, and Glossary.

word is '**archetype**'.[55] Archetype's a key, most rational *idea*. Archetypes embody informative not energetic causation. Attractors/ archetypes, the source of natural order, purpose and coherence, are of two kinds found in both upper and, filed in memory, lower mind.[56]

individual	*general*
somewhere	*nowhere/ everywhere*
local	*non-local*
factual	*conceptual*

The conscious attractor, causal or potential mind, is identified as Archetype; this Primary Archetype is Logical; it is Single, First Cause or *Logos*.[57]

The lower, unconscious attractors, causal or potential matter, are identified as archetypal memories or multiple, fixed forms in universal mind. In the human case these secondary, passive records are reflected by fixed patterns in your own subconscious mind; they are memories known, in part, as instinct. The way such morphological attractants, that is, archetypes are linked to polar, cosmic infrastructure is an issue Chapter 13 will elaborate. Gradually we'll pin the crucial, causal notion down.

Polarity/ Duality

The second existential principle is *polarity*. Poles express the fundamental duality inherent in things as opposed to its lack in the Monopole, Nothing. Indeed, cosmos can be seen as tensions anchored between poles of various sorts; nature, physical and psychological, is seen as a dynamic interplay of opposing and yet complementary forces. **But the fundamental polarity is between Essence and existence, between Potential and action or, in other words, Nothing and things.**

One yet many. Light, for example, breaks into spectral bands. Most of these, from extreme intensity to extreme long-wave cannot be seen by human eye. We can see visible light but no other band. If the universe is hierarchically banded then what is at one vibratory level is not at another but may still exist in the same 'space' simultaneously. For example, are not myriad radio and TV broadcasts passing through you at this second?

Certainly, bureaucracies are tiered. Orders issue from on high. The question is the nature of that highness. Is the source of command material or not? An obvious immateriality is, as an aspect of informative potential, mental creativity; and mind's essential substance

---

[55] *SAS* Chapters 7, 8, 16, 17 and 19; also Glossary and Index.
[56] see *figs*. 7.2 and 7.3 levels 2 and 4.
[57] *SAS* Chapter 5: Top Teleology.

is its consciousness. Do any physical phenomena consciously behave? Or is mind a separate 'wave-band'; is it an element in its own right and, therefore, physically nothing? In which case what bull, except materialism's papal writ, is issued to forbid consideration of, as well as universal forms of energy, universal types of information in the form of universal mind?

In short, is the Source of Universal Spectrum immaterial? Is the font of every something None? Where something comes of nothing you are in a world of paradox. This, precisely, is our world.

Check *fig.* 8.1. Its legend shows that Immaterial Being is identified with Consciousness. If Consciousness beyond the relativity of mind and matter is Top Absolute then obviously its diametric opposite must be the special case of consciousness - none. Non-consciousness. This is the state of our material universe.

Such polarity is generally expressed as follows:

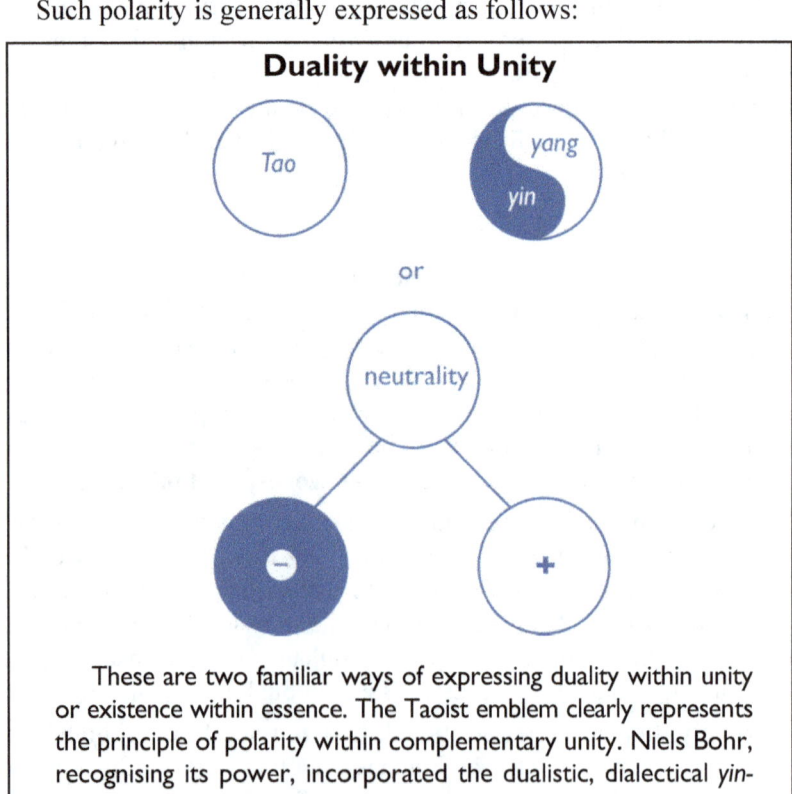

These are two familiar ways of expressing duality within unity or existence within essence. The Taoist emblem clearly represents the principle of polarity within complementary unity. Niels Bohr, recognising its power, incorporated the dualistic, dialectical *yin-yang* design into his family crest! But the mystic quest, Taoist or otherwise, is to transcend the duality of existence.

fig. 10.1

Duality within Unity. We noted (Chapter 8) that existence is binary yet contained within Essence. Transcendent Essence is nothing existential but, nonetheless, the cause of existence. From an Absolute Viewpoint (called Enlightenment) all is, within its Creative Projection, quintessentially one. However, while this Perspective knows that All is One, relative perspectives see things differently or, as in matter's case (oblivion), doesn't see at all.

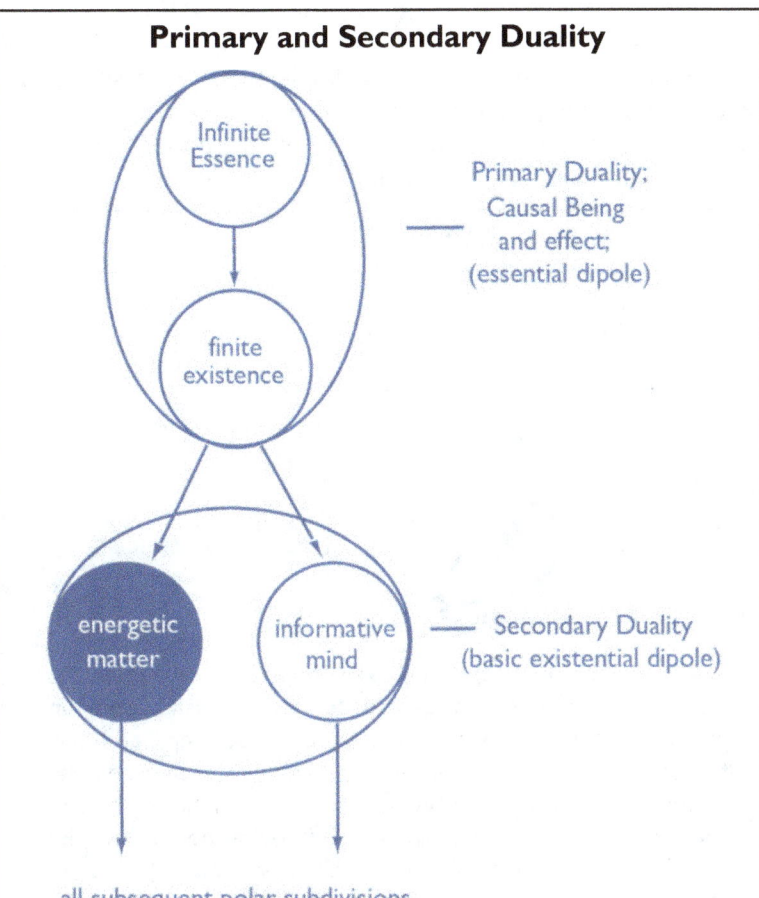

The philosophical term 'monism' means that you stipulate a single, fundamental and universal quality or entity. 'Dualism' means that you stipulate two.

Saturated materialists are 'monists' whose single element is energy/ matter.

On the other hand, Natural Dialectic proposes 'Essential Monism' with 'existential dualism'. The single, infinite 'substance' is identified as The Concentrate of Pure Consciousness. Motion of this essence gives rise to the dualism of existence; the basis of this dualism is information/ energy or mind/ matter. The couple is hierarchical in that the

> former generates the latter; immaterial generates material; informant mind generates (in a way that Natural Dialectic describes) informed patterns of energy. Finally, just as there exist universal matter and specific biological forms, so there exist both universal mind and individual instances of mind.
>
> **fig. 10.2**

If we think in terms of cosmic gradient, scale or spectrum that extends from pole to pole we add an element first introduced in Chapter 8 - the Dialectical Inversion. It is reflected in the species explored in that Chapter's 'Source and sink' and 'Voids'.

Primary Inversion

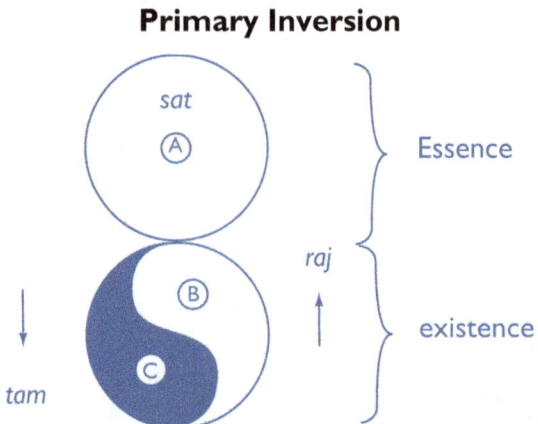

It comes out opposite. **An inversion means that something turns out opposite its origin.** Cosmic poles and inversions sound complicated but can easily be simplified using a 'flipped open' version of the yin-yang symbol. You can see that A (Unity) is the antithesis of B/C (duality); also, within existence, B (metaphysical mind) counterpoints C (physical body). It is clear that the diametrical or 'far opposite' of A (Essence) is C (the physical universe); in other words the extreme inversion of Essence is material solidity and *vice versa*. Such antitheses are regarded not so much as mirror images but *inversions* of each other. 'As above, so below' runs the reflective maxim. Dependent B and C reflect the nature of independent A but, in the process of realising A's potential, this reflection also involves an inversion. It involves the 'organic' kind of development that expresses 'inside information' outwardly. According to its logic Inmost Essence (A) devolves a polar creation whose final stage ends up as non-conscious matter (C). **This, the drop from consciousness to non-consciousness, is the primary, creative inversion.**

fig. 10.3 *(see also figs. 10.1 and 10.2)*

The drop from consciousness to non-consciousness is the primary, creative inversion. Agents of this switch are anti-parallels, the antagonistic fundamentals[58] called *raj* (action ↑) and *tam (*fall ↓).

Their action is expressed in a simple link from physics. *This will (in Chapter 11 and, as previously, in Chapters 4 and 6) describe the three basic conditions of information and of energy as (sat) potential prior to (raj) action and (tam) exhaustion.*

Electrostatic charge accumulates in purple clouds. From motionless potential builds the storm. It breaks and pours and, spent, is over. There is calm before and after cloudburst but of very different and inverted kinds. Is not exhaustion the reverse of powerful potential? From potency to impotence, before is not the same as after.

So with the cosmic storm we call existence. **A gradient slopes from potent to impotent peace; a balance swings from poised until inertial equilibrium**. Thus, as day's lease gives way to night, light is blackened. Start to finish takes the course of an inversion. The immaterial pole of consciousness is twisted in descent to an extreme, non-conscious, material constriction - perhaps in the form, at base-pole subtendence, of a black hole. In terms of Primary and Secondary Dialectic, pole to anti-pole inversion is registered as an *innate* 'switch' from Primary Right to secondary left. It reflects the general fact that a (*tam*) left-hand extreme represents the negative opposite of a (*sat*) characteristic - even though they may appear superficially similar and the same word, such as 'rest', may be misleadingly used to describe them both. Their true characters are diametrically apart.

At the final stage of 'organic inversion' consciousness has been turned inside out and fallen, therefore, into the non-conscious physic of bodies animate or lifeless. Natural Dialectic would suggest it has been twisted, upside down and out, into the bailiwick of physics and its chemistry. Conscious concentrate has been completely lost; what started out without non-conscious energy has turned to nothing else. It has become a plenitude of lifelessness. If oblivious matter constitutes the last flight of creation's stairs its final step is turned to stone. The scale from zone of soul to that of pure, non-conscious energy completes the opposition of its poles that Natural Dialectic calls, we've seen, reflection by asymmetry. *This is what, in universal terms, we mean by paradox of inversion.* **Our starry universe is clustered round the lowest rungs of a Great Ladder, the last step-down out of a higher, grander cause - an outcome that completes inversion of, as well as you,**[59] **the body of our cosmos.**

However, the Dialectic also *resolves* polarity in the form of a third, central component, Balance.

---

[58] Chapter 4.
[59] see *SAS* Chapter 17: Caduceus esp. *fig.* 17.4.

equilibria	*Equilibrium*
expression	*Potential*
↓ *inertial equilibrium*	*dynamic equilibrium* ↑
impotence	*power*
loss of bounce	*cyclical wave-form*
flat-lining/ death	*homeostatic vibration*

Fig. 3.3 showed an inversion from Potential to impotence; and the last stack of Chapter 8 (the drop from Inaction to inaction) anticipated the current example, Equilibrium to equilibrium.[60] Balance, peace, equilibrium. (*Sat*) Equilibrium is poise that precedes action; it is the source whence a creation or an action is derived. Of the vectored pair (*raj*) dynamic equilibrium is seen as carrier of energy, cyclical wave-form or homeostatic vibrations such as stabilise life. And (*tam*) inertial equilibrium represents the locked, finished condition of capture and terminal loss of creative ability. Such triplex scale is mirrored in creation.

Informative potential (mind) exists with energetic yet non-conscious, automatic matter. Each is a major aspect of underlying Essence. Two-in-one is also one-in-two. The framework thereby, as we've seen, complements existential relativity with Essential Absolution. Thus its stacks represent Essence (Top Pole) and Essential Characteristics set, top to toe, against a range of vectored relativity we call the existential field. **This existential matrix, composed of informed and informant (energy and information), is called the conscio-material gradient.** From its Top, 'Strong' Pole of Subjectivity this 'sliding scale' falls to the special case of 'weakness' - a base pole of none, that is, of totally oblivious objectivity. In other words, at the top informative mind is dominant, at base informed but automatic energies.

This nether base, non-consciousness, is creation's body; such objectified embodiment of Essence we know as our spacious, sunny universe. At the bottom of its states of matter rests solidity.

---

[60] Glossary: equilibrium.

Chapter 11: Elements of the Basic Existential Dipole

*The basic elements of the existential dipole are **information** and **energy**. From these co-principals creation is derived. They compose the basic coefficients of existence.*

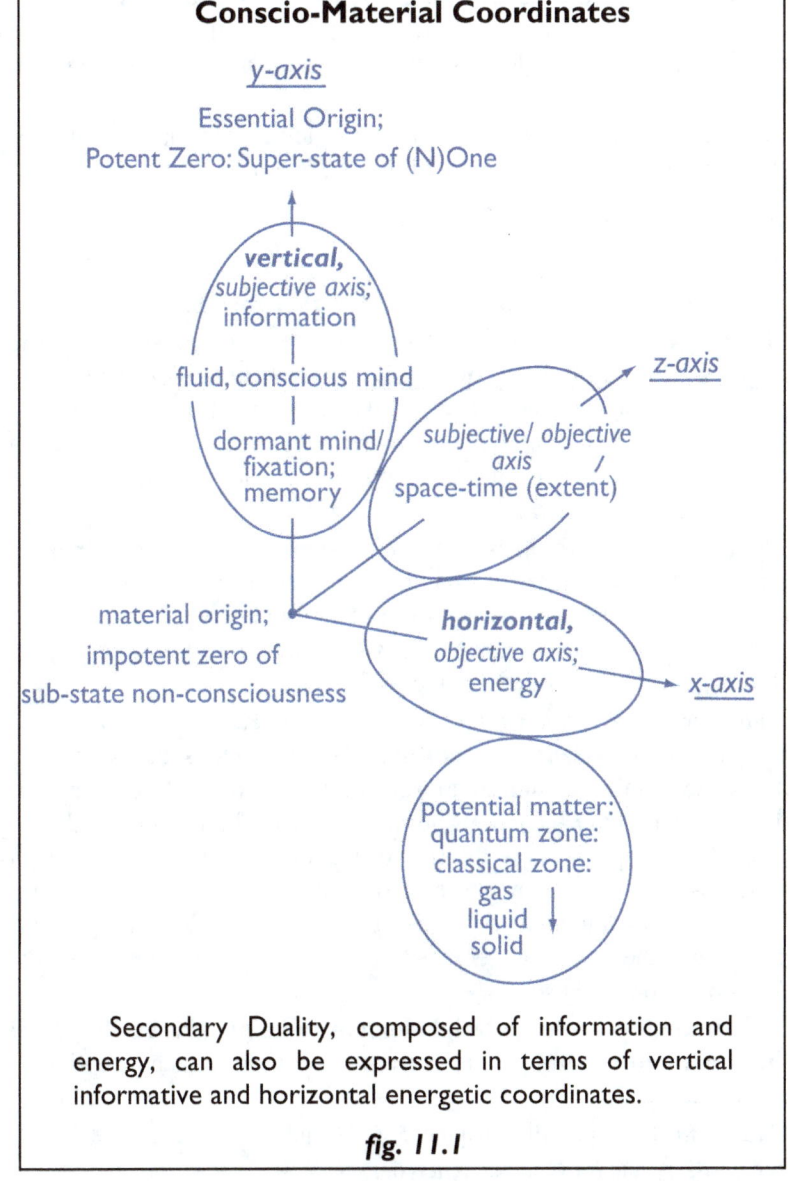

fig. 11.1

Their duality can be graphically expressed in terms of subjective and objective components.

↓ *object*	*subject* ↑
material energy	*immaterial mind*
physical	*psychological*
entropic tendency	*negentropic tendency*

These are, respectively, mind and non-conscious energy. The basic tendencies of this pair are, respectively, negentropic and entropic. Informative negentropy is the natural tendency of inquisitive mind; it also inhabits the nature of specific, organised creations as opposed to energetic entropy's aimless indifference, breakdown and general, physical decline.

As illustrated in *figs.* 7.1 to 7.3 the cosmic conscio-material gradient includes a sub-gradient of mind and, lower, one of matter.

Upper Pole - Information[61]

You may say there's only matter.

I say there are material *and* immaterial elements.

If more than physic does exist then what is metaphysical? **Natural Dialectic's immaterial element is information**. Information is arrangement by or for intent. Objects never carry words or numbers. Only minds describe, count or assign them patterns. 'Unnatural' objects, such as an image or machine, are informed by mind. Could anything inform the objects of our natural world?[62]

***Bottom-up*, everything is seen as energetic interactions. Energy's the physical informant.** Information's therefore, even in the case of brain, evolved by chance and natural law. Oblivious physic generates, in its complexity, myriad 'off-chances'. Is this, materialism's 'metaphysic', an illusion, a confusion or a doctrine of the truth?

***Top-down*,** *signal information is the gift of mind and never matter.* **It is irreducible to scientific scrutiny. Yet, as we'll see, its semiotic metaphysic dominates debate about creation and our lives.** Pick up a postcard, menu, letter - anything informing you. The object is reducible to chemistry and physics but the *sign* or *message* it conveys is not. Signs and signals always have a purpose; objects never do. In other words, information's fundamental to our being but does not fall, in a semantic sense, within the scientific remit. *Particularly, mindless origin of bio-code is an irrational hypothesis.*

Information is an immaterial element in its own right. Why? **It is nothing without mind.** Mind conceives and understands conceptions. It

---

[61] *SAS passim* but especially Chapters 5, 6, 16 and 19.
[62] see Glossary and *SAS* Index: archetype.

is active and commits ideas into material orders that are called, in the Dialectic, passive information. Informant and informed. Mozart's mind, through instruments, informs material sound. Is not the real heart of energy through air you hear as music therefore immaterial? Is its primary arrangement psychological or physical? Is it physical or metaphysical?

Buy Mozart's music! CDs and floppy-discs are materials carrying information. Have you ever weighed one accurately then rubbed it clean? Deleted all the information? What, when you reweighed it, was that information's mass? Zero? None at all? Do the same to your computer. The essence of computers is their programs and their database. Delete this essence and peripheral hardware weighs the same. Because informative capacity inhabits the *arrangement* of material - not just any old arrangement but one with meaning that serves purpose or accords with principle. Purpose weighs no more than understanding; both weigh just as much as meaning. In the scientific balance meaning weighs as much as abstract theory. Each is lighter than a feather. Weigh every thought in every mind that ever was; weigh universal mind itself. Such aggregation measures zero grams and so the scales tell nothing; nor do the finest registers of time and space as much as twitch. If, though, information's absent massively, what is it? If immaterial then, as this volume shows, it renders atheism (but not science) most illogical! Indeed, Norbert Wiener, mathematician and founder of cybernetics and information theory, said *"Information is information, neither matter nor energy. Any materialism which disregards this will not survive one day."*

There is, furthermore, known neither law nor process nor sequence of events through which oblivious matter can create or collect information. The latter is not a property of matter. Purely material processes, unguided except by natural law, are fundamentally precluded as *sources* of information. Information is not a thing itself but a representation of physical things and metaphysical entities.

Information's hierarchical as well.

Hierarchical Information

Active (Creating) Information and Passive (Created) Information

Information is phrased in terms of awareness. **Of the major cosmic sub-divisions mind is termed active information. Although it may exhibit highly active behaviours matter is, with respect to information, non-conscious; it is passive, automated and in this respect creatively impotent.**

Clear distinction needs be drawn between informative mind and informed matter.

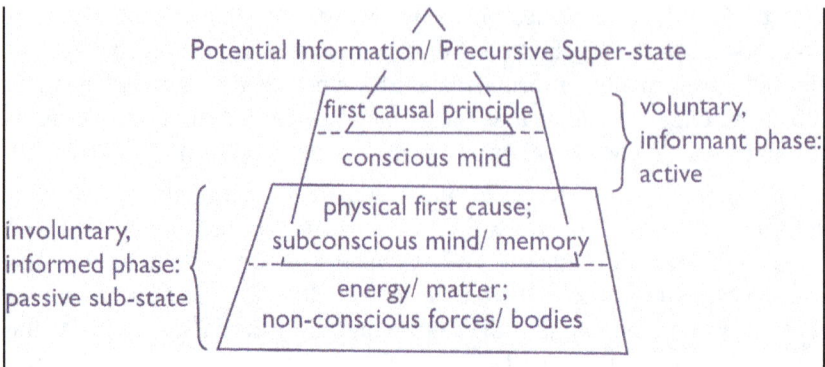

There also exists, however, an active, informant phase of matter - energy; and a passive, informed phase of mind - a sub-division called sub-consciousness. Sub-consciousness is mind's storage facility. *Memory is the natural recorder of finished events in active, conscious mind. This lowest level of mind also constitutes, in dialectical terms, the point of psychosomatic linkage with matter. In other words, passive mind in the form of memory becomes the transcendent, causal principle that orders matter.* This hierarchical view of memory is not standard, is not materialistic but, we shall see, fits well with psychological, biological and physical facts.

Opposing Hierarchies

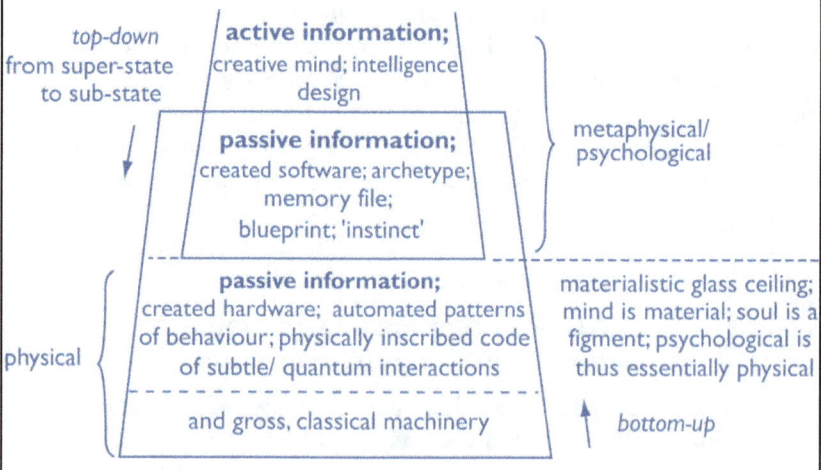

Top-down and *bottom-up* views of the origin of information are here juxtaposed. *Top-down*, information originates in active mind; *bottom-up* it starts with passive matter. Since information is the source and sustenance of life which view makes better sense?

> In the *bottom-up* view passive leads to active information. Matter/ energy emerges from nothing or is, perhaps, eternal. This matter fortuitously, in a process called abiogenesis, creates a cell coded by *DNA*; such a physical carrier of code itself generates information; and such informative, though random, chemistry somehow eventually makes consciousness. Intelligence. *Hardware creates software.*
>
> The *top-down* view is absolutely the reverse. Information is the immaterial gift of mind. Just as software is composed of non-material instructions that control the hardware, so a distinct immaterial element, conscious mind, designs software instructions (stored in the files of memory as archetypes). Such metaphysical programs may be reflected in material carriers. Similarly, sub-atomic particles can be conceived as carriers of archetypal code; the behaviour of non-conscious substance is thus metaphysically controlled. *Software defines hardware.*
>
> **Finally, within existence active information involves awareness: passive information does not.** Passive information is fixed, automatic, reflex. The domains of its storage are subconscious mind and non-conscious matter.
>
> *fig. 11.2*

We can summarise the gradient of Mount Universe as conscio-material. This gradient involves a dipole (*figs.* 10.3 and 11.1). Of these the second pole, matter, is a special case of the first, mind. Its case is zero. That is to say, material energy in every form is involves total absence of immaterial factor. It is subjectively passive, that is, non-conscious. It is, in subjective terms, never informant and always informed; but in objective terms it may, as far as an observer is concerned,

below	*Top*
subsequent order	*Archetype*
manifestation	*Latency*
its excitation	*Immaterial Field*
its expression	*Informative Potential*
consequences	*Initiator*
↓ *outer/ external*	*inner/ internal* ↑
passive mode	*active mode*
objective aspect	*subjective aspect*
non-conscious/ physical	*psychological*
passive forms of info.	*active/ semantic forms of info.*
informed energy/ matter	*informant mind*

automatically assume both voices. Oblivious but energetic matter may actively inform and be passively informed by other objects or events.

Firstly, according with cosmic fundamentals, we take a look at the triplex nature of information.

(Sat) Potential Information

Potential precedes any possible action or result. It is a prerequisite or precondition. The cosmic precondition is Unconditional Essence. It is the Super-State Transcendence of Chapter 9.[63]

(Raj) Active Information

Active information[64] is only created in conscious, choice-flexible mind. Your conscious mind is an information exchange. It interprets signals from its exterior; it is informed by subconscious instinct and sensible circumstance. Equally, it is informant; it purposes to understand and actively arrange its world. This activity involves ideas, executive design and devising the semantic expression of control.

Informative potential is, for sure, not physical; the metaphysical informant issues orders from a centre of control. In this case, might universal mind somehow instruct the specific habits that nature automatically repeats? If so, what kind of archetypal entity comprises the potential matter from whose presence issue, automatically, mindless behaviours we call rules - the rigid laws of nature?

(Tam) Passive Information

Informant causes; what is formed is informed. Its specific behaviour is passively, automatically received.

Thus, passive information[65] **is the expression, external to conscious, flexible mind, of active information.** Such information may be dynamically exchanged (as in the case of speech or body language). Otherwise its impressions are stored either in subconscious mind (featuring relatively inflexible memory, instinct, archetype and so on) or using matter (where instruction is carried by arrangement on chemicals such as clay, papyrus, ink, *DNA* or other messengers). Storage may be fixed (as in a file or photograph) or dynamic (as in a running film, program or automated mechanism).

---

[63] see also Chapter 10: Causality, Chapter 14 and *SAS* Chapter 5: (*Sat*) Potential or Transcendent Information and Top Teleology. One common analogy of Archetype is as a prism from which are diffracted the colours (or wavelengths or levels) of mind; another is of a Basic Vibration (*Om*) from which the various sounds of cosmic opera derive.

[64] see also next Chapter, Chapter 13 and *SAS* Chapters 6 and 14.

[65] see also next Chapter, Chapter 13 and *SAS* Chapters 6, 15, 16 and 17.

Passive information includes grammatical rules, syntax and the objects (or units) of their construction - words that compose linguistic or other natural code. A code is an agreed set of symbols arranged to format information. Such **upper linguistic/ codified level** involves particles (say, letters of an alphabet); forces that regulate their conjunction (punctuation); grammar (say, the elements of a language such as noun, verb and so on); and syntax. Syntax is the convention or legal framework within which symbols are ordered; its law naturally determines those structures allowed and those not. Thus the *upper syntactic level* acts as a filter through which order is communicated to and from the *lower (environmental, statistical or quantitative) level* of data items - physical phenomena. Inward, subtle, immaterial regulation orders outward, gross, material expression. **Seen thus the laws of nature are, in type, linguistic code.** Written in particles and forces on blank sheets of space, cosmos is a dynamic book. It is thus easy to understand (where *logos* means speech, word or order) why the ideal, causal level of cosmic oration might be called *Logos*. The universe might indeed be logical.

This is because creation issues from within; active information orders passive. Mind generates code. **Matter can act as a vehicle or storage medium for data but never, being subjectively impotent, generate code and therefore coded information.**

In other words, information has meaning to a mind; such meaning (or semantic) may be organised into passive, symbolic receptacles such as language. But, while the concept of 'message' cannot be expressed in terms of the solely descriptive concepts of physics and chemistry, at the lowest, physical level it *can* be expressed through vocal, written or other material form. This is the **lower linguistic/ codified level**, that of physical expression. Sound waves, paper, clay or *DNA* can register ideas; so do machines or artefacts of any kind. Thereby the constraint of code is locally translated; and, with respect to natural archetype, the general is turned physically specific everywhere; that is to say, the lower pole, material cosmos, orderly appears.

Lower Pole - Energy

Matter is energy.

An argument has, however, been enjoined viz. that natural cosmos is composed of energy (including matter) *and* immaterial information. If 'natural' is equated with 'material' then 'unnatural' information has no natural cause. Non-conscious, physical phenomena create neither codes nor, which codes carry, messages; although intrinsically devoid of meaning or purpose they may, however, *convey* both of these. Things may be dynamically or otherwise arranged, as in the case of music, language or machine, to express meaning; matter may, to its molecular core, be shaped by the various logics of purpose.

How could this apply to subjectively inert, non-conscious matter?

Natural Dialectic would propose that the cause of physical phenomena is metaphysical. There exists, as well as non-conscious physical energy, a level of metaphysical mind - a 'field of passive information'.[66] This 'file' of universal memories is identified as archetype.[67] In the order of creation, from inward immaterial causes to outward, material effects, such immaterial information bank constitutes the link between mind and matter. Metaphysical archetype precedes physical outcome.

Secondly, we take a look at the triplex nature of energy or, if you like, of physics; and we begin with the above-mentioned archetypal linkage called potential energy (or matter).

(Sat) Potential Energy

Potential precedes possible action. It is a prerequisite or precondition for results; but Natural Dialectic does not think of physical potential in the way that physics does. Potential matter is, as we've seen, a metaphysical, informative affair.[68]

Informative archetype (potential energy/ matter) is thought of as letters in a universal alphabet, as bits, bytes and routines of a computer program or, better (since vibrations/ wavelengths correlate with forces and energies), notes whose harmonics compose the cosmic opera.[69]

(Raj) Active Energy

Active energy,[70] appearing in accordance as these notes are energetically plucked, is the physical product of informative potential.

The space and time of mind are not the same as matter's. From non-local, omnipresent archetype are issued the automatically, mindless behaviours we call rules - the rigid laws of nature.

These behaviours are harmonic oscillations. They manifest locally and are known to physics as the quantum world of particles and forces; they are the minuscule and energetic base from which, as icebergs in a sea of

---

[66] *figs.* 7.3, 8.2 and 11.1-3; also Glossary and Index: archetype. This level is also called material super-state, transcendent matter or potential matter.

[67] see previous footnote; also Index: mnemone, typical.

[68] Chapters 9-11; see also *SAS* Glossaryand Index: archetype.

[69] see *SAS* Chapters 6: Music; 11: Nature's Holy Ghost; 16: Psychosomasis and How Does the Connection Work?; also Index: music, archetype, mnemone and harmonic oscillation.

[70] see *fig.* 7.3 - matter-in-principle; also *SAS* Chapters 7-12 for a discussion of the physics of both matter-in-principle and practice. Physics is, of course, the study of energetic interactions of precise, pre-set and wholly automatic kind; and, put simply from a dialectical point of view, chemistry observes the push-and-pull of polar charge.

energy, gross sensible matter is formed. Natural Dialectic calls this level of creation, the quantum phase, subtle matter or **matter-in-principle**.

(Tam) Passive Energy

Passive energy is 'locked'. It is known to Natural Dialectic as the classical phase of gross matter or **matter-in-practice**; and it forms the 'bulk' level of obvious, sensible events. The constituency of its objects is 'locked', 'bonded' or relatively 'fixed'. Visible, microscopic and suspected but invisible chemical phenomena were, until the late 19th century, the only ones counted by mankind. The classical boundary with matter-in-principle (quantum phase) might be drawn at the level of atom.

If the linkage between potential matter and matter-in-principle is, presently, unknown and ignored, the link between matters in principle and practice is unknown but far from ignored.

We have two theories to explain the physical world - relativity (which cannot say how subatomic elements work) and quantum mechanics (which ignores gravity). Both are accurate but they contradict each other. In other words, the description of matter-in-principle contradicts that of matter-in-practice. **Our gross sensible reality is not what it seems. There is quantum reality beneath or (in terms of Mount Universe) above it.** In this respect, we say that quantum is more 'inward' or 'towards creative centre' than classical.

In this case a proposed linkage between above and below is a theoretical structure called quantum gravity.

We are happy to tolerate quantum gravity because it can be reasonably, mathematically described. However, unless fundamental, abstract equations supplied by science data books amount to its description, this is not the case with the higher, invisible link of archetype. After all, archetypal memory is not, unless perceived in terms of mathematics, consistent with materialism's mind-set.

Who, though, ever measured thoughts, dreams, experience or subjectivity - which certainly exist - mathematically? Why, therefore, should we reject an attempt to nail down the link between potential matter and its inferior, matter-in-principle? This is what Natural Dialectic calls the psychosomatic interface - an interface between lower mind and its externalisation as the basics of physics, chemistry and biology.[71]

It is clear by now that, beyond scientific vision, a whole new world inside creation opens up. This, ignored by materialism and, methodologically, by science, is its immaterial interior.

---

[71] *SAS* Chapters 6: Information's Infrastructure - Code; 7: Precondition/ Potential; 16: Psychosomasis and 19:The Basis of Biology is Information and Conceptual Biology.

Informed Energy

A couple of diagrams will help revise what we have learnt about energy and information, the co-principals of creation.

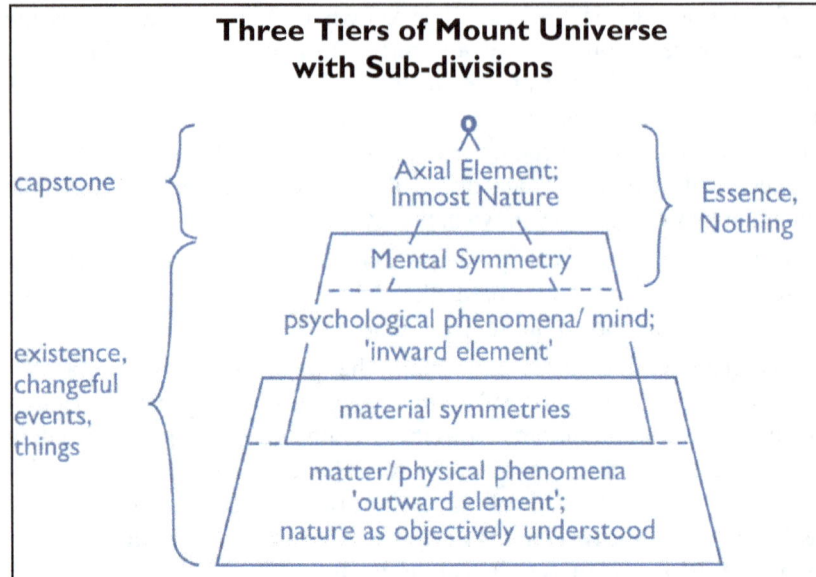

Major sub-divisions of cosmos are (*Sat*) Essence, (*raj*) mind and (*tam*) matter. The information in these ziggurats will be elaborated throughout the course of the book.

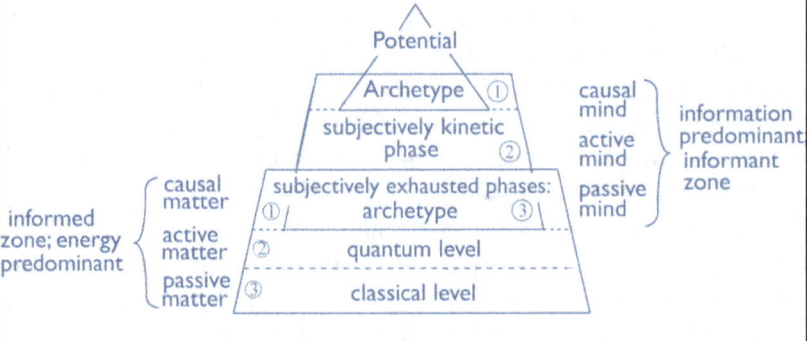

① pre-active plan, 'seed' or causal archetype
② internal informant, pattern-maker; primary effect
③ external structure, fixity of pattern; secondary effect

Within each major division minor (numbered) sub-divisions can be drawn. These numbers are related to the three cosmic fundamentals. For mind these are (*sat*) causal superconscious, (*raj*) conscious and (*tam*) passive sub-conscious levels; and for non-conscious matter a psychosomatic link grade, (*sat*) causal/ potential matter, (*raj*) quantum matter-in-principle and (*tam*)

> bulk or bonded matter-in-practice. This final, minor sub-division itself involves a familiar sub-sub-division into gas, liquid and the final, most fixed and 'static' expression of them all, solid.
>
> It is often claimed that distinction drawn between action-at-a-distance forces, quantum systems and those of large-scale, apparently divided-up objects and events is artificial. This is because quantum effects, although less obvious on the large-scale, exist at all levels. A similar proposal might be made with respect to potential matter and its archetypal effects.
>
> **fig. 11.3**

Objective energy, the stuff of physics, is about action; subjective information is about the motion of will, power of purpose, meaning and control. Together they comprise plan and its realisation.

In summary, to inform means 'to give form, shape, intelligence or organising power'. It is to communicate a pattern of behaviour. Psychologically, informative patterns involve sense, knowledge, understanding and desire; they include the attributes of meaning and purpose. The province of such knowledge is 'active information' - conscious mind; an obvious example of actively informed energy is a machine. And the natural province of 'recorded and therefore passive information' is memory - subconscious mind. Subconscious mind, including instinct, is classed as 'passive information'. The automatic characters of energy are also passively informed. They behave by what is known as natural law and show as patterns of an object or as courses of events. Any pattern of behaviour is explained by its cause; its cause is its reason; and this reason includes both innate capacity and the external permutations of circumstance. Reasonable behaviour needs a plan; information is a deed's potential; and mind precedes material behaviour. *What, however, is the cause of physical behavior, the reason behind natural law?* **For now, it's simply noted that if the way material things relate cannot be generated by those same relations then, logically, transcendent laws of nature must be metaphysical.** The provision of initial conditions for all natural events means that their origin is super-natural. Physical cosmos is a *carrier* of information but is not its cause.

What, therefore, does *fig.* 11.3 imply? **It implies that the universe does not have a single dimension. It has three - ultimate, original potential and its dual expression, mind and matter.**[72] **The quality of events within these grades depends on the proportions in which different types of information and energy are involved with each other**.

It equally implies an order of expression, a sequence down the

---

[72] also *figs.* 7.1-3, 9.1, 10.2 and 11.2.

scale of creation. Mind over matter; from mind mindless matter. *From a materialistic point of view this is outrageous.* It commits a capital thought-offence. Off with its head! Truncate the universe from immaterial information! No doubt, the view that matter is subordinate to mind is, from a human point of view, against our intuition; that conscious power is implicit in creating the explicit bulk of mountains seems preposterous. Let us, therefore, at this moment re-clarify. There is not only aimless matter in this world; there are meaning and intention too. The association of information with energy should not be seen as a Cartesian split but as a gradual, mutual, proportional involvement. *We are seeing, as this book unfolds, how the game transpires; how, from the larger picture's point of view, do things work out?*

While creation's immaterial interior is less obviously present in the physical exteriors of matters-in-principle and practice, the reflection of archetypal order is clearly apparent in the conscio-material case wherein coordinates of both information and energy (*fig.* 11.1) play a complex part - life on earth. **The very basis of biology is informant code**. And it is against our intuition, strongly and reasonably, that complex codes occur at all by chance. If the basis of specifically codified and functionally operative biological forms is information then the basis of their material construction is immaterial.[73]

Indeed, as in the case of all purpose-built physical machines, cell-based bio-form is irreducible to physics and chemistry alone. Mind immaterially supplements. Its presence is behind the purposes, such as the survival-driven and thus reasonable behaviours of every bio-mechanical part of every integrated whole. Where in the universe do lifeless gases, stars or crystals demonstrate intention, tenacity of tendency and cleave to coded, specified complexity by many means determined to 'live on'? Mechanisms and machines show rationale; they are obviously energy informed.

Life is certainly built on information; it is developed from symbols that encode purpose in the form of programs. Such information is carried, for its ink and paper, on a complex molecule called *DNA*. *Therefore, we have to ask again, what is information? Where does it come from? Can unguided, natural forces generate the message that is needed to construct the simplest cell? In other words, in the beginning was there information or not?* **Materialism says no; holism, in a logically formulated way, includes its immaterial element.**[74] **It, therefore, says yes.** *Non-conscious energy is, physically and biophysically, archetypally informed.*

---

[73] see also Chapter 15: *passim.*
[74] *SAS passim*; especially Chapters 5-6 and, as regards life, 19-25.

Chapter 12: An Act of Creation

Information is the boss. Matter is creation's automatic slave: it is, like potter's clay, the unquestioning worker that is realised as potted order. An idea becomes material creation's pot.

We tend to use the word idea to denote some artistic, technological or adventurous novelty. Actually, mundane ideas (called desires) are continually driving us. After the last chapter, however, we are in a position to understand how ideas and the creative process that occur in us, as microcosms, might reflect the macrocosmic, universal dynamic.[75]

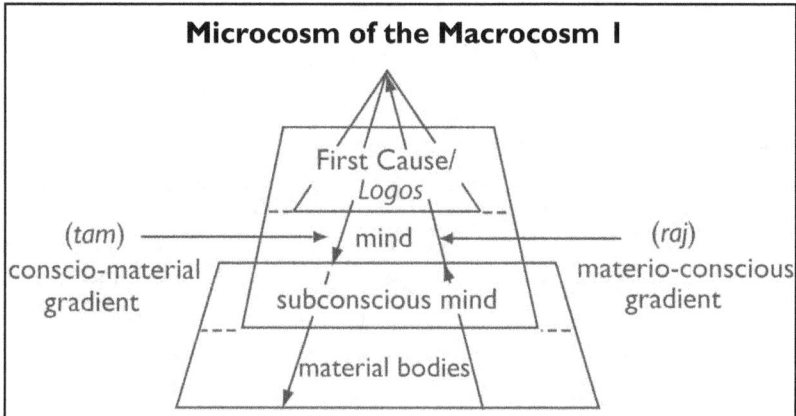

Natural Dialectic represents the creation as a ziggurat or concentric circles but what is its 'dynamic'? What is the light of the dialectical rainbow? Check *figs*. 3.1 and 7.1 to reconstruct your idea of gradient and vectors. Two factors apply - inward, informant mind and outward, informed energy. This holistic, conscio-energetic couple compose creation's dynamic. The subjective fraction is 'volitio-attractive', that is, a combination of will-power (pushy 'I will do') and desire ('I want' pulling); and the objective remainder is the push-and-pull of physical energy.

In this case 'the creation express' runs down a consciomaterial gradient; its expression is a form of fiat - 'let it be'. First Cause or *Logos* is a concentrate or super-state of consciousness. From an original, that is to say hierarchically superior, exercise of will devolve creative principles that inform non-conscious practice, that is, the energetic components of material creation.

---

[75] see also *figs*. 7.4 and 13.1; and *SAS* Chapters 13-26.

> Between the concentrates of consciousness and non-consciousness (*Logos* and material bodies) each vector of a 'polar dynamic' has its part to play. Expressions of (*tam*) gravity of materialisation and (*raj*) levity of dissolution oscillate in all psychological and physical events. **Local loss or gain of energy or information describes change in the universe.**
>
> fig. 12.1

Could your own constructions mirror cosmos? If so, how does the spark of an idea develop into plan and, orderly, its material execution?

Orderly Creation

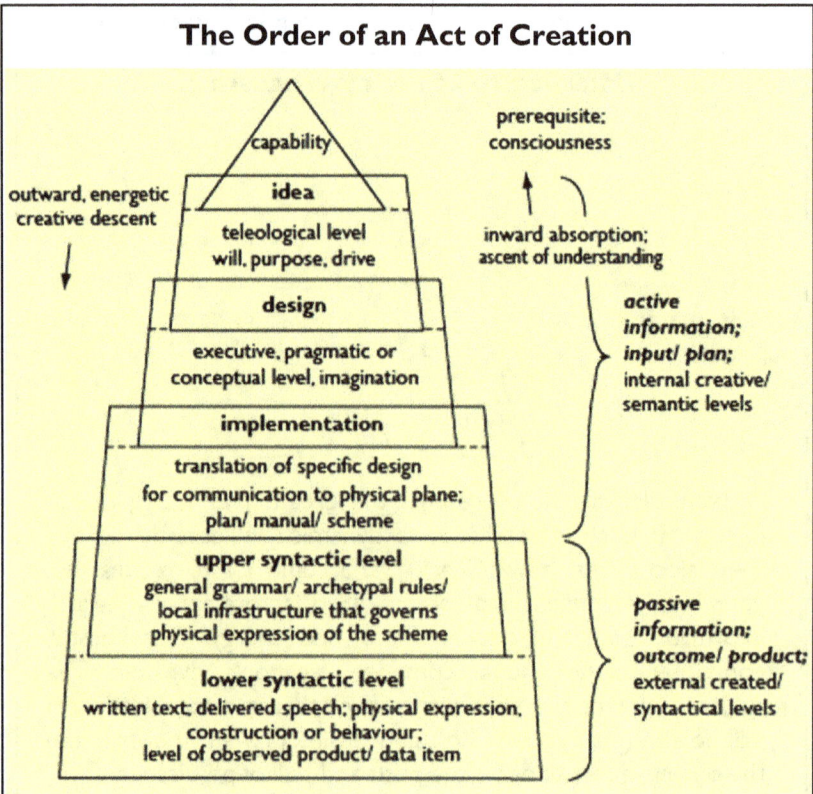

In this hierarchy levels are conceived of as nested but co-existent. A creative act develops, organically, like a seed to flower. The act itself 'descends' from metaphysical to physical levels - from mind to body. Such 'descent' may occur very rapidly as, for example, in a conversation; or more slowly in, say, working out the detailed solution to a technological problem. Acts of creation range from trivial through important to cosmic in scope.

> Conversations, solutions to problems and so on also involve 'ascent'. They involve an act of 'counter-creation, comprehension or understanding of what is going on. This applies with full force at top, teleological level where principle, motivation and purpose are grasped. When you have got the idea you have the reason.
>
> Not all stages are necessarily engaged. For example, a person ordered in an inflexible, military way to do something scarcely enters the top three levels; instinctual and reflex information loops are passive.
>
> Have you ever 'got the idea'? Grasped what someone or something actually meant? What about the Cosmic Idea? Since, in this view, the Origin of cosmos is alive then a full comprehension of the Idea amounts to Communion with its 'thinker' - Pure Subjectivity. Such Essential, *Top-Down* Subjectivity's Perspective, transcendent, independent and yet commanding all existence, is therefore of The Very Purest Objectivity as well. The two are one. Knowledge (Science) is their Super-Nature.
>
> Although irrational according to the tenets of materialism Illuminative Knowledge is, of course, the product of experimental technique. The mystic, as was Christ or Buddha, meditates. If he completes the cosmic loop and his hypothesis is proven right, what a result! What, Pascal might add, a winning bet, a vindication!

fig. 12.2

Fig. 12.2 indicates that an act of creation amounts to phased intent. It is, simply, the materialisation of an idea. A simple idea is, in execution, elaborated. You work out what to do and then, according to your plan, do it. If creation is the re-arrangement of the world according to one's wishes, you and I remake it all the time. The major phases of intent are comprised of active and of passive information.

Sometimes, no doubt, interactions and constructions ramify. Complexity occurs; and of complexities there are, at root, two kinds.

The first, <u>non-purposive</u>, is a function of energy transformations. It involves the confinement of energy (e.g. in sub-atomic particles) or loss/ gain of energy in changes of state (e.g. from formless gas to such differentiated variation-on-theme as snowflakes show or *vice versa*). In aggregation or in action nature's physic is mindless. Its <u>complexity</u> (whatever its beauty in the subjective eye of a beholder) is <u>passively derived</u> from the automatic behaviours of energy informed by rigid 'rules'. In this sense matter is a 'no-innovation zone'; having no freedom of will it is lifeless and automatic. Scientific research and application focuses on this indiscriminate kind of complexity.

The question is whether you can infer a purpose behind the 'design' of apparently non-purposeful complexity. Could the universe be, by anti-chance not chance, a product of design? Could it be both wonderfully and purposefully made? The response from scientific materialism is that 'there's no necessity to think that way because it's all necessity - the name for natural law'. Its answer is, in other words, a resounding, study-stopping 'no chance, since it's all (including the origin of natural law) by chance'.

Where physical possibilities are precisely or, at least, statistically measureable, psychological ones are more flexible. The second, _purposive_ kind of complexity works the other way. It is a function of information gain, expansion of consciousness and a capacity to grasp and purposely, creatively exploit the principles and possibilities inherent in any circumstance. An increasingly concentrated focus of attention wakens to greater capacity, flexibility and possibilities for specifically ordered, coherent or _active complexity_. At each level of ascent the degree of coherence may improve; the degree of ingenuity, adaptability and innovative complexity may increase. For example, human mind and its society constantly respond, adapt and build new, elaborate structures to combat the age-old exigencies of a problematic life on earth.

Active, purposeful complexity works against the 'downward' wear and tear of time and chance; it codifies and specifies design - which chance cannot. Purpose is the driver of informative negentropy. The more specifically detailed a construction the higher its negentropy. *It is natural that such construction (e.g. a machine or work of art) exhibit maximum negentropy at its finalised plan; and also that, while archetypal plan is not impaired, its physical expression is subject to entropic wear and tear. Such decay leads inexorably, despite the possibility of negentropic reordering (repair), to eventual destruction.*

As with machines so with cosmos and, it is submitted, its initial creation; and as with cosmos so with humans and the informative potential of all other kinds of biological plan. Is not gradual genetic degradation of the physical projection what science finds?

In short, informative negentropy specifies and codifies design. It is the fundamental instrument of biological survival, intellectual enquiry, technology and artistic creation. We continually experience it. Its proof, in artefacts and actions, pervades our lives.

Law and Order

Laws order. Information and order are closely allied. Let's, therefore, check the *active* order of creation in terms of the upper dipole of existence, information.

From a *bottom-up* point of view natural order is paramount. It is described in terms of material laws of nature based on equations whose constraints are also dubbed 'necessity'. Conscious mind and its species of order is, though powerful, simply a contingency. It is an evolutionary 'extra' whose recent, unexplained emergence on the scene is cosmically irrelevant.

A deduction that omits mind from the world equation until the very least and last precisely reverses the *top-down*, dialectical order of information. The latter's order of creation runs, first and foremost, from mind. Things start in mind. The order runs from psychological to physical, from active to passive information, from creativity to its creation. ***Top-down* hierarchy emphasises an order of information that runs from active (mind) to passive (matter).**

Psychological is linked, as we've just seen, to the physical expression of Cosmic Dialectic through a psychosomatic (mind/body) phase we know as the residence of 'fixed thought' - subconscious memory. Therefore, where memory is the natural, paperless retainer of ideas, it is also the depository of 'filed plan' called archetype. Potential is an unexpressed capacity. Archetype is energy's precursor; it is the 'template' or the 'field' that, activated, will accordingly express orderly mental or material creation's play. Potential in this sense is causal. Archetypes *cause* orderly succession; what succeeds or 'comes below' is dependent on their regulation. And, simply put, the Archetype of mind's psychology is both informative potential and its activator rolled in one; we call this 'primal action' *Logos*. And the causal archetype of physic we have called potential matter; it is the databank of our material universe. Archetypal template amounts to principle, to preconditioning or law and order in position prior to when the action starts. Law and order rule, in every part, the cosmic game.

Let's now turn and ask what constitutes an act of creation?[76] *What is its order?*

Are you, creator of creations, not a faint but noticeable simulator of the way that great creation's system works? On a conscio-material gradient passive follows active information; input precedes output. Sensation starts in matter and ends up in mind but creative action starts, *top-down*, in mind and ends specifically re-arranging bodies in the world outside. It 'drops' from causal reason towards a reasonable effect. Let us now see how the last chapter's triplex subdivisions of the basic existential coefficients, information and energy, are expressed in mind and body as you yourself elaborate and physically realise ideas. Firstly, therefore, follows an account of the 'internal' or 'subjective' aspects of an information loop. These are conscious teleological, pragmatic and semantic phases.

---

[76] *SAS* Chapter 6.

Capability

This is the basic ground or, in Dialectical terms, the peak from which creation drops. It is Essence to existence. It is Source or precondition we call (*Sat*) *Potential Information*. Such initial is metaphysical. The precondition from whose context an idea is formed is metaphysical - unless in all respects you think ideas are physical!

Purposely Down to Earth

(*Raj*) *Active Information* is only created in the great processor, conscious mind. First issues an **idea**. This is the potential, seed or genesis of any active, formulating process. There accompanies intention to elaborate the embryonic inspiration's promise. Such expansion's motive power is will-power and desire. *Indeed, will and desire are (as noted in fig. 12.1) the psychological equivalent of physical electromagnetic radiation.* Will is like electricity, desire like magnetism. Together they are the light and life of mind; the '*volitio-attractive*' charge of their attention draws, fires and bonds. Such current lights up life. This process, grand or trivial in scope, is the basis of all purposes.

Getting Your Way - Pragmatics

Mind's dynamo streams from focus of attention, a concentration of interest; and interest may intend to order things or events according to its own purposes. Call this **executive design**. Pragmatics, the internal process of consideration, is set at the level of executive or systems analyst. *Transmission of purpose through the pragmatic level involves, in the outward direction, imagining and planning its realisation; and, on the inward, deciphering another's plan.*

Desire expects results. *This is active teleology.* Optimally, it involves manipulating options, working out exactly what is wanted of the communication or design, what its format or construction will be and the effect it will have.

What Do You Mean?

Causal reason generates meaning. All active information is meaningful but may or may not require externalisation. In other words, the thinker may or may not want to communicate his experience. If he does he sends a message and will want a relevant response. The meaning will need to be externalised and, so that its external receiver understands, it needs to be specifically arranged according to mutually agreed code.[77]

Such recipient will be animate. In this case, a message involves

---

[77] see Glossary; also *SAS* Index; alphabet, code, language (bio- and cosmological).

conceptual interpretation. On the other hand, interaction with inanimate, non-conscious material may result in the creation of an implement, machine or some impact on the environment; and, passive on passive, reflex materials continually interact. In such communication an incommunicative material side is always called the 'passively informed recipient'.

Implementation of a plan involves semantics whose specific meaning transcends the generality of grammar and syntax. The latter are simply vehicles of reasonable expression; their immaterial symbols are needed to make connection with the material world and thereby order it. They translate mind to matter. Coding and decoding, using speech-through-air, written word or other forms of signal, are core semantic business.

To this fundamental end mind employs symbols, often words, to express both pragmatic and meaningful intent. It specifically arranges them to execute a command, a wish or an intention. Printed, electronic or spoken signal carries the message of its creator. Paper, ink or electrons are physical entities, meaningless *per se*. However, their commander has controlled their order to provide a channel of communication. **A command erases randomness; it is a deliberate restriction imposed to cause a non-random outcome.** It employs an agent of restriction - sign, symbol, code of one sort or another. *This is the world of signals, semantics. What is conveyed is not a matter of chance.* **Whatever is encoded is intentional.** Code and chance are chalk and cheese. They never mix. Language, other symbols and the construction or decipherment of meaningful communication (called semantics) make up the third level of information. In other words, this is the level at which ideas are framed in symbolic code or blueprint before their presentation in material form.

↓ *irrationality*	*logic* ↑
no-message	*message*
no-code	*code*
physico-chemical maelstrom	*psychological scheme*
chance/ accident	*teleology/ purpose*
aimless construction	*architecture*
mindlessness	*mindfulness*

Such assertion is important for understanding an orderly universe, especially codified biological forms. **To affirm: code is always the result of a mental process.** *A coder takes no chance. Randomness is eliminated.* A compiler is a mindless mechanism but a programmer is not. He determines the code and its operation: error is rigorously debugged. By definition, mistake or randomness degrades information; and the job of any editor is to eliminate interference, 'noise' or mistake. **Chance neither** creates nor transmits information. On the contrary, accidents always (unless accidentally reversing a previous degradation) degrade meaning and, by degree, render information unintelligible.

In short, symbolism and simulation are mind's agencies. They are its metaphysical intelligence. They make sense. Although *meaning* is the important, active ingredient behind codes, data transmission and storage, nevertheless these passive instruments of communication are important. We need frameworks (hardware) within which to manage information (software).

Information's Infrastructure - Code

(Tam) Passive Information involves the agent of expression, code. A code (as opposed its manipulator) is an agreed set of symbols arranged to format information. As already noted in chapter 11, such **upper linguistic/ codified level** involves particles (say, letters of an alphabet); forces that regulate their conjunction (punctuation); grammar (say, the elements of a language such as noun, verb and so on); and syntax. Syntax is the convention or legal framework within which symbols are ordered; its law, allied with a vocabulary, naturally determines those structures allowed and those not. Thus the **upper syntactic level** acts as a filter through which order is communicated to and from the lower (environmental, statistical or quantitative) level of data items - physical phenomena. Speech, using vocalisation, air and ear, is a physical phenomenon; so is writing on support material.

Musical, vocal and artistic expressions all employ some learned abstraction. Is mathematics perhaps the most abstract code of all? The linguistic, harmonic and mathematical rules are learnt and stored in memory and thence, as required, retrieved for use. In each case subtle, inward, immaterial regulation orders outward, gross, material expression. The message always comes in code; and where there are personal purposes (as in the case of conscious animals) there will be variable expression. The ability to devise or learn different, flexible languages is the function of a framework laid in animal instinct; the general capacity to engage in orderly, flexible communication is innate. Man, the information hunter, is capable of enormous yet most orderly flexibility and therefore great subtlety of expression.

If a basic code is found in any system, you might conclude that the system originated from a mental concept, not from chance. You might therefore conclude it had an intelligent source - especially if that code is optimised according to such criteria as ease and accuracy of transmission, maximum storage density and efficiency of carriage (such as electrical, chemical, magnetic, olfactory, on paper, on tape, broadcast etc.) to its recipient; and if, above all, it works and orderly instructions are unerringly responded to.

However, although codes are simulated on physical materials their primary storage location is metaphysical. It is in subconscious mind whose other name is memory.[78] And, to reiterate, the name of memory in universal mind is archetype. Seen thus the laws of nature are, in archetype, linguistic code. Written in particles and forces on blank sheets of space, cosmos is an open, active book. It is thus easy to comprehend (where *logos* means speech, word or order) why the ideal, causal level of cosmic oration might be called *Logos*. The universe might indeed be logical. The question is how archetype works inside out and thus enables the material universe.[79]

Computer programs, musical scores and architectural plans are, as well as written words, examples of codified expression. And perhaps the nearest human forms to natural language are vibratory music or mathematical abstraction. At any rate, we call a natural code (as opposed to the communications of biological forms) **archetype**. In this view passive information correlates with matter in a universal way. Informant archetype shapes the character of particles and forces. Thence, energy on energy, predefined behaviours shape all material objects and events. In this sense, the cosmic lower pole (non-conscious energy) consists of passive information.

Such basic linkage, information with matter, will be elaborated throughout the rest of the book. It therefore bears emphasis that, in terms of Natural Dialectic, the latency of cosmic language is always present and, wherever roused to action, expresses an orderly creation. The *upper syntactic level* of information's infrastructure, code, is equated with **potential matter;**[80] and such matter is, in turn, equated with **archetypal memory, archetype or, simply, nature's memory.** Such automatic regulation of behaviour appears, behind events, as natural law.

No doubt such equation, potential matter with archetypal memory, is heterodox; what else, by definition, can a fresh suggestion be? The issue is, when immaterial information is included in the deal, whether such equation smoothly integrates within a post-materialistic paradigm.

The Lowest, Physical Level

On a conscio-material gradient passive follows active information; input precedes output. Creation issues from within; active information orders passive. Mind generates code. Code's informative potential conditions both action and outcome. Matter can act as a vehicle or storage medium for data but never, being subjectively impotent, generate code and therefore coded information.

---

[78] *figs.* 7.2, 8.2, 11.2 and 12.1; also *SAS* Chapters 15 and 16.
[79] *SAS* Chapter 16: How Does the Connection Work? And Index: harmonic oscillation.
[80] *figs.* 7.1-3 and 12.2 Upper Syntactic Level.

In fact, whenever an intention is physically expressed it works through the technology of mechanisms and machines. Machines, which include biological bodies, operate according to physical law and fulfil their function using, in one form or another, energy. They specifically accord with plan and inform the world in ways unguided nature can't. As such they obviously link information (that is metaphysical) with energy (that's not). An example is the way our brains operate vocal chords to the order of a particular wish.

Thus, at the lowest, physical level we express information through vocal, written or other material form. This is the **lower linguistic/ codified level**, that of physical expression.

Therefore, bump! This is the part that bottoms out in light, sound, fluid patterns and in crystalline solidity. This is the 'external', 'objective' or physical side of cosmos - the (*tam*) exhaustion of original potential. Here, at this lowest and entirely non-conscious level, the automatic play of matter reigns supreme. And, while forces and atomic particles comprise its primary expression, the hard, bulk edges of the universe that we survey is its secondary. *At this lower syntactic level we find data items, that is, materials on which arrangement is intrinsically imposed.* The phase appears as an object, event or reflexive pattern of behaviour without subjective quality, context or intrinsic meaning. It is the level of raw data *per se*, that is, objects in whatever form they appear *before* higher inspection, manipulation or interpretation. No subjectivity and only husk-like objectivity remains. The values inspection may impose on such oblivious objectivity are often numerical and its description measures 'thing-ness' alone. In short, the lowest level of 'information', devoid of sense or meaning, is vested in force or aggregate of atoms, that is, in physical being.

We can now realise that, although the oblivious and automatic dynamics of external phenomena show no intent their matter's aimlessness is internally coded. **The behaviour of the world is guided from within.** Such codified guidance is especially obvious in the very complex, obviously programmed and purposefully developed forms of life.[81]

In short, code is devised and stored in mind but information's physical expression obviously employs material arrangements. **The alphabets of such code are bio-chemicals (supremely, *DNA*), binary nervous code and the sub-atomic elements, forces and atoms of physics and chemistry.** The former couple are specific to life forms - possibly, at least with respect to *DNA*, anywhere life may exist in cosmos. The latter also constitute a universal code. This code dictates, through the agents that a study of phenomena elucidates, the way things naturally turn out. *Looked at this way could the universe itself express a covert plan? Could cosmos be, at root, a Grand, Dynamic Text, a form of Opera, a Great Idea?*

---

[81] Chapter 15; also *SAS* Chapters 19-25 and *A&E* Chapters 8 and 9.

Chapter 13: Triplex Psychology

Triplex Mind

In the order of creation (and of our behaviours) information governs energy.[82] Mind - metaphysical, knowingly informant and informed - hierarchically precedes physical matter. *Therefore the next three chapters will run from mind to matter (or energy) and their very highly informed combination in biological operations.*

On the analogy of an information spectrum or a conscio-material gradient mind has many levels; but, for purpose of simplicity, we divide

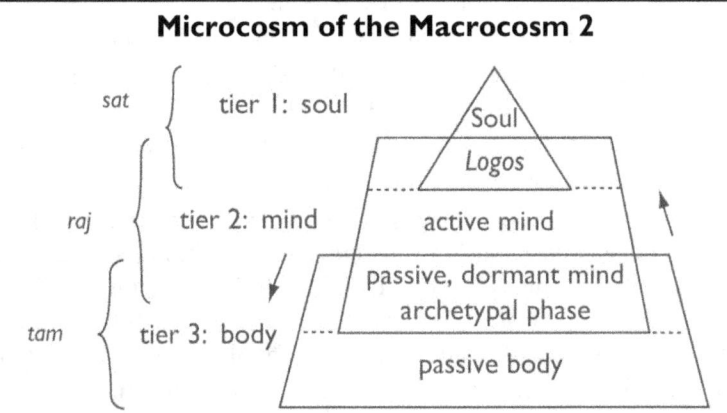

Do you reflect creation as a whole? Is your constitution in the image of a three-tiered universe? In this analogy the anti-parallels run *in* (for understanding), *out* (for action's creativity); input, output - their two tracks lead to and from an information-centre at the head. In this respect they form an image representing you. This is how, in body's tier-3, passive zone, your nervous system is arranged; *indeed, as we'll see, it represents the informative way that your whole body is disposed, that is, it represents the structure of biology.*

At this point it may also be noted that, from a *top-down* point of view, created 'mind-with-body' is seen as a 'soul vessel'. Thus every different pot (or organism) contains the same pure water. In this manner a human may be seen as soul having a physical experience rather than a body erroneously imagining its metaphysical soul.

fig. 13.1 (see also figs. 3.1, 13.3 and 13.4)

---

[82] Chapters 11 and 12.

tier 3/ tier 2	Tier 1
body/ mind	Psyche/ Soul
lesser selves	Transcendent Self
conscio-material range	Conciousness
↓ physic	metaphysic ↑
energetic	informative
material	immaterial
non-conscious	states/ grades of mind
body/ tier 3	mind/ tier 2

it into a 'visible light' band of mundane human consciousness ranged between possibly much broader, higher (UV) and also lower, instinctive and subconscious (IR) bands. Such 'triplex mind' reflects the triplex levels of information described in Chapters 11 and 12. **_The three tiers are, basically, superconscious, conscious and subconscious_. They compose a scale of subjectivity**.

Microcosm of the macrocosm - a simple, three-tiered cosmic ziggurat represents the overview.

The word *'psychology'* means study of *psyche* (from a Greek word generally translated 'soul'). But current 'psychology' were better termed *'noology'*. This is because the discipline studies mind, for which the Greek is *noos* (from the verb 'to notice' or 'to think'), and not *psyche*. Or does it? Perhaps not even 'noology' goes far enough. **Scientific materialism can, by definition, only allow physical composition.** Thus contemporary research swings even further towards a hands-on, neurological commitment.

But is thought an object you can analyse, like nerves, underneath a microscope? If the key element of mind, your subjectivity, has simply been reduced to ghostliness and the central, immaterial dimension of psychology's reality electrified then how far past a neurological glass-ceiling has science scaled?

The Neurological Delusion

Could mind live apart from brain? Dead bodies aren't, for sure, alive; a mindless corpse would then imply that, in a universe of matter, immaterial mind exists and constitutes the core of our terrestrial experience.

Such substitution, metaphysical for physical, is, however, scientifically 'unacceptable'. 'Impossible' - whatever future studies might disclose. Death is just 'parts failure', a dislocation of molecular mechanics whence once arose the corpse's mind. Don't informative computers also terminally crash? Won't you? 'Immaterial' is a word too far; it breaks the mind-set's basic rules; it kicks a prop that's critical to sustenance of naturalistic faith. So, creed decrees, thought's entirely a nervous matter. *Thus, naturally, psychology emerges as the study of neurological phenomena.*[83] Well-educated and evolved professors tilt their weights towards neuroscience as the guru. Carbon, oxygen and more - soul is the activity (incredibly, incomprehensibly complex, mind you) of particles. Consciousness 'emerges' from non-conscious molecules grouped in some special way. Isn't thinking generated by the soft-wired workings of a brain? Just sling sufficient atoms, in the form of nerves, together - they'll become no less than self-aware!

That mind is a physical illusion is the neurological delusion. To mistake its neural correlate (as, say, registered in brain scans) for experience itself is an error as basic as taking the electronic pulses in a wire for all there is to telephonic conversation. It is a prime, elementary fault, a first category philosophical mistake; it might be termed full-blown, psychological mythology. To identify consciousness as an illusion is itself, denying the reality of one's own experience, a pernicious - even dangerous - delusion. Who, however, cares for error when the guesses of materialistic faith cap all?

Does Brain Originate or Mediate?

One party, it is clear, believes life is a phantom of the atoms. Brain *causes* subjectivity. Thought (therefore belief and all the purposive effects of will and faith) is part and parcel of nerve chemistry. And what is the *experience* of consciousness? The essentially robotic view of neuroscience holds that nerves *are* consciousness. We just don't yet understand, the faithful purr, how brain's 'emergent properties' can squeeze experience out or how the juice that's 'you' must be exuded from its molecules. A revelation is, however, prophesied. Materialism's scientific certainty decrees that life will be reduced to chemistry and mind experimentally identified as simply due to complicated ionicity! You are a product of your physiology and so, at root, your genes alone. Life has, hasn't it, to be an electronic after-thought?

Nervous particles and atoms aren't, like atoms anywhere, alive. Therefore, if life is made of them it shouldn't be alive! *Thus the other line suggests, conversely, that brain isn't an originator but a mediator.* A filter. A sophisticated interlink beween mind, body and the latter's physical universe.

If so, it is a **transducer device** that, like any mediation network (e.g. radio), must be sufficiently well-constructed to handle large volumes of two-way traffic.[84] It accepts environmental signals and translates them (↑) 'upward' into mind's experience; and issues orders (↓) 'downward' into body chemistry. As an organ of 'cockpit control' its 'dashboard' accurately connects an immaterial mind to a material body and, thereby, physical

---

[83] *SAS* Chapter 0: Scientific Delusions; also Chapters 5, 6, 13 and 18.
[84] *SAS* Chapters 0: Opposite Directions of Mind and 16: Signal Translation.

conditions. Of course, young pilots (babies) have to learn to fly; thence we and other kinds of creature navigate, in the vehicle of body, various sagas on the senseless stage of matter. In this view mind and brain, although compounded, are quite separate entities - the former metaphysical and latter physical. Brain chemistry's identified as a design that expedites exchange of information. Your head is thus a medium!

Making no material difference by adding immateriality, the Dialectic simply reconstructs creation on the basis of a 'conscio-material' duality. In short, perhaps brain neither does nor ever did enjoy a seamless, subjective experience. The implications of this seminal idea are so extensive that this whole book is exploring them.

Consciousness

This is what it's all about. Without it you are nothing. The star of every play is mind; the kingpin of psychology is consciousness. What is the 'thing'?[85]

For Natural Dialectic there's an immaterial element of information. This element of knowledge isn't physical. Non-conscious matter is a special case of its subjective absence. It is *pure non-consciousness*. Gases, streams and solid bodies don't know anything. Their oblivion's polar opposite is total wakefulness. Of what, you ask, does this consist? As matter's pure non-consciousness exists could not a concentrate of immaterial information - pure consciousness - have being too? So that creation's root turns out to be oblivion's antipode - *Potential Knowledge, Latent Field of Knowing, Pure Consciousness.*

Materialism doesn't like this sort of phraseology at all. *Top-down*, if non-conscious matter forms the cosmic sink its source is consciousness. *Uncreated Consciousness is primary and existential forms are secondary.* **What light is to physics Essential Transcendence is to psychology; in other words, the only absolute measure of consciousness, and therefore psychology, is Transcendent Super-Consciousness.** From mind holistic distillation therefore reaches high; but materialism recognises only base degree, the lifeless one.

Now, therefore, let us sketch the *top-down* form of an essential psychology.

Top-down Psychology

Top-down, any attempt to reconcile science and psychology must logically start at the top (Conscious Soul) rather than bottom (matter). It must start at Axis, Centre, Source or Pivot rather than with the effects of subsequent informative or energetic motions. Therefore we deal with the

---

[85] *SAS* Chapters 5: (*Sat*) Potential or Transcendent Information; and 13: Consciousness.

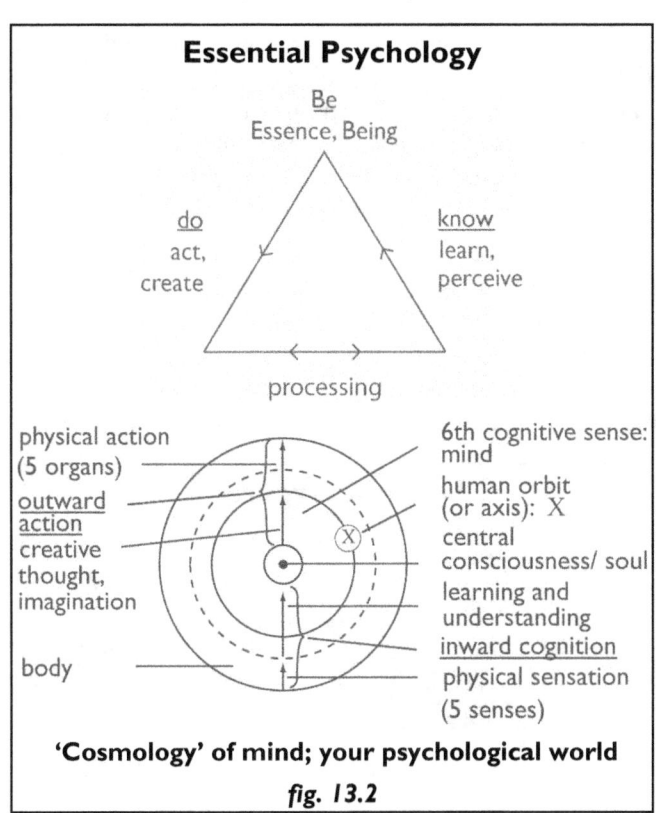

fig. 13.2 'Cosmology' of mind; your psychological world

do/ know	Be
info. out/ info. in	Information Centre
subordinate principles	First Principle
logical expression	Primal Latency
objective/ subjective facts	Subjective First Cause
body/ mind	Psyche or Soul
passive/ active expressions	Potential
↓ to do	to know ↑
info. out/ output	info. in/ input
non-conscious/ physical	metaphysical/ conscious
reflex	voluntary
differentiation	integration
creation	knowledge
behaviour/ response	understanding
actor/ action	perceptor/ perception
outward focus	inward focus
down from centre	up to centre
externalisation/ mind to matter	internalisation/ matter to mind

two conscious states, super-state transcendent and our ordinary, restricted awareness. Then we fall to cover, at the sub-conscious end of mental balance, the conditions of dormant mind, dreaming and deep-sleep; finally we inspect the psychosomatic domain of instinct, personal memory and archetype. The latter account for the psychosomatic connection between mind and its sub-state, non-conscious material body.

The grades of cosmic hierarchy are not set in space. They are vectored in accord with energetic power and comprehensive scope of information and control. *Top-down*, hierarchical psychology explains the levels and directions to which operation of the mind is geared; and, with oscillations in between these levels, nervous correlation in the human brain. In which gears does your life predominantly run?

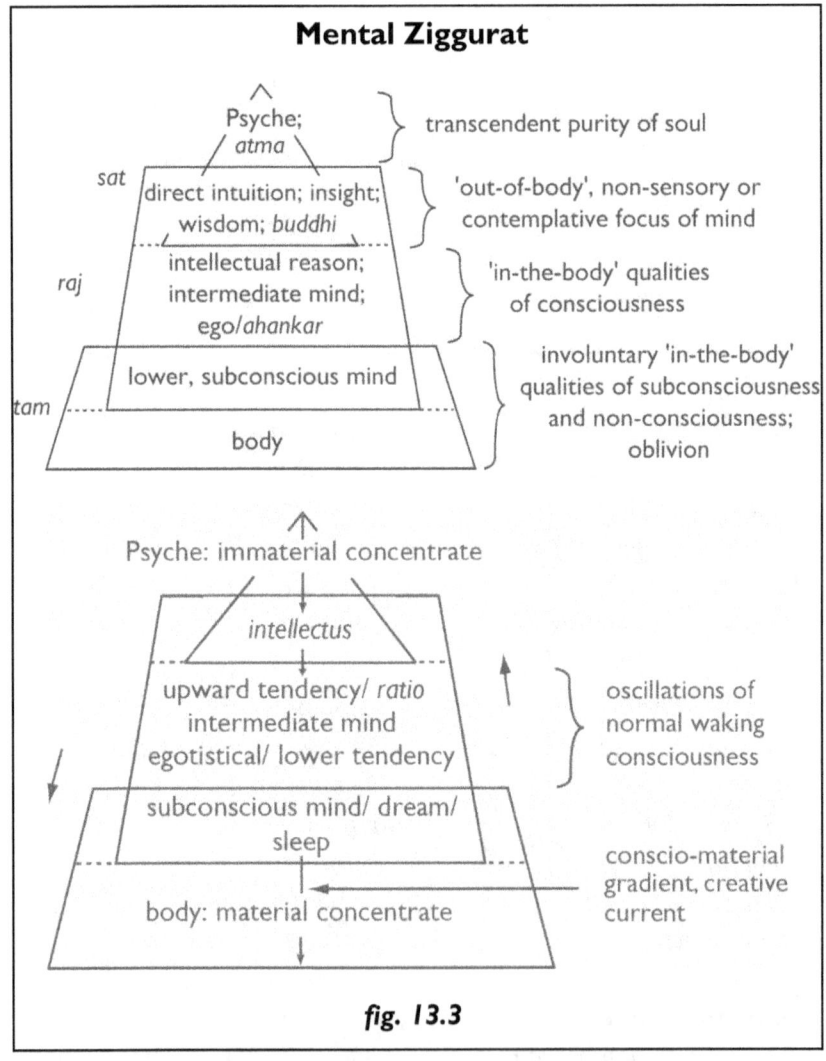

fig. 13.3

First State of (Super-)Consciousness -
The Psychology of Transcendence

Life's real secret rests, according to the zealous atheists who in 1953 discovered how it coiled, in a chemical called *DNA*. In their view mystic truth is actually a 'God Delusion' that's evolved from genes, their proteins and electric fields of charge.

Life's *real* secret is, according to the mystics, something we've already met - Informative Top Teleology.[86] It is an immaterial moment of enlightenment. Nanak analysed no followers' genes but was his Central Truth a figment of uneducated lies? Is Buddhist enlightenment, *Nirvana*, an illusion? Did Christ miss Crick's real secret? What is Lao Tzu's Alpha Moment, Ramakrishna's Great Idea, the world's Great Love?

Materialism stutters to explain Tzu's Moment. Its reflex is denial. It discredits, charges with deceit or plain explains it right away. Erase all immateriality! Yet what is love but living union? Sexual union is physical; friendship is dependent; but Friendship, metaphysical communion with Single, Living, Cosmic Source, is a mystic's holy grail. This, the 'Unifying Theory of Metaphysics', is realised as Knowledge, Truth and Love. **It is, therefore, no exaggeration whatsoever to say that human civilisation is constructed from and around a supra-religious tryst with the eternal moment. Human faith, hope and ideals are logically derived from the materially meaningless experience of transcendence.**

Second State of Consciousness -
The Psychology of Waking Normalcy

	created cosmos	*Transcendent Potential*
	restrictions of awareness	*Awareness*
	lesser selves	*(No-)Self*
	lesser truths/ shadow games	*Clear Truth/ Reason*
	orbits	*Axis*
↓	*negative/ no*	*positive/ yes* ↑
	lower/ more reflex	*higher/ freer*
	passive/ involuntary info.	*active/ voluntary info.*
	effect/ outcome	*cause/ input*
	less conscious	*more conscious*
	sub-rational/ instinctive	*calculating/ reasoning*
	more instinct	*more thought*
	necessity/ non-will	*conditioned free will*
	puppet/ body self	*thinker/ mental self*

---

[86] Chapter 11: (*Sat*) Potential Information; also *SAS* Chapters 1 *figs.* 1.1-3 and 2.1; and 5: Top Teleology

If the first state of consciousness is absorbed with Top, Essential Principal, the next two involve intellectual understanding, manipulations, feelings and the pragmatic business of negotiating life's sub-principles (such as healthy life-style, cultural regulations and apprenticeship); and, not least, dealing with the satisfaction of its needs and physical desires.

Embodied mind is our condition and, as such, mind and body interact. Our state is subject to the vagaries of body and its circumstance. But is this temporary state the only one?[87]

Higher Normality involves inward, contemplative focus. Such focus tends (↑) 'upwards' towards a growing comprehension of general or universal principles. In seeking release from apparent disarray, confusion and limitation-in-egotistical-forms such 'mind-in-principle' best approaches and therefore reflects the characteristics of its highest grades - *Archetypal Logos* and Transcendence. These traits include, as well as unification, continuity, coherence, integration, relationship and communication. *Such positive, right-hand principles educe a well-balanced, focused and attractive personality. Such a side is idealistic and strives for wisdom, beauty and love.* In dialectical terms, an increase in the predominance of (*raj* ↑) right-hand characteristics '*expands*' consciousness.

The next band is *Intermediate Normality*. Ego is self's intellectual executive and weaves, from sense to thought and back, with serial in-out logic to achieve its ends. As a wilful, manipulative schemer 'mind-in-practicality' is a driving instrument of problem-solving and achieving goals. Like any middleman, however, *ego* has to balance both sides of equations and in so doing cuts both ways. Its centaur-like operation may involve higher (↑) or lower (↓) moods, tendencies and desires; and it employs, at different times and in various persons, varying focus of intelligence.

Lower Normality. If there's an upside there's a (↓) downside too. Now comes a dark day. The other side of mind is a *separator*. The main downward principle of our psychology is *confinement* with its correlates of structural discontinuity, differentiation and individuality. *Negative, left-hand principles educe disruptive, distracted and unattractive expression.* The body-focused isolate is selfish, restricted, aggressive, turbulent and as demanding as unpopular. The lowest cast of conscious mind shows inertia, depression, laziness and stultification. A human in such common but '*constricted*' consciousness is 'animal' inasmuch as an animal is most involved with unreasoning, sense-driven

---

[87] As the band of visible light is to the electromagnetic spectrum so the range of the human band of mind, our normal peep-hole, is in the whole conscio-material scale of creation. As regards the transcience of this peep-hole see *SAS* Chapter 18: Death.

reflex. Continually noisy sensation is consumed by lusts and physical survival; there 'bubbles', in addition, a continual, diffuse chatter, daydreaming sometimes called 'default mode'. 'Lower mind' lacks focus; its negativity seethes with passionate attachments or else with fear, hate and jealousy. Its base-pole ignorance is dimness and a deadweight drag.

The Third State of (Sub-)Consciousness

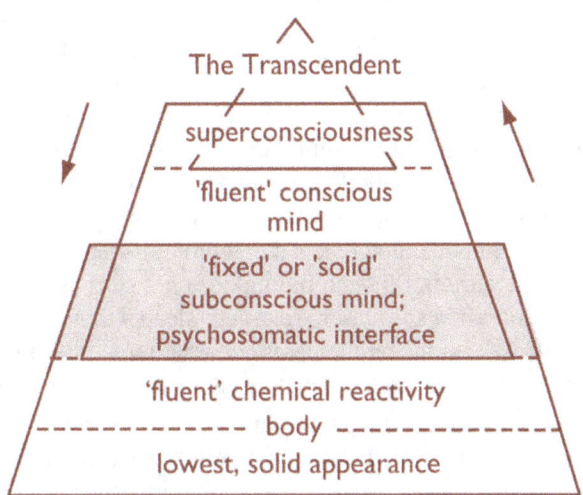

fig. 13.4

Are you ready for a fall to the diffuse conclusion of psychological entropy, for a subjective drop into the labyrinth of underworld?[88] You know what it is to be mentally as well as physically exhausted. You've often dropped off into mind's flat, dark condition we might call inertial equilibrium. 'Little death' is not the world's end so let us take a snooze cruise; it is time to fall asleep.

The Psychology of Dreaming

For dreamers dreams are real enough; but the experience is untrammeled by either external events or the ability to reason. Waves wash equally on what is in their path; a torch shone randomly around picks out disconnected or illogically connected objects and events. The files are scattered, narrative is blurred.

In this apparent chaos is there reason? Dreams serve, like waking mind but lacking sensory restraint, to relieve circumstantial pressures in the form of dangers, problems and anxieties. Subjective equilibration is the game. For relief's solution, relaxation and release of tension, dreams

---

[88] *SAS* Chapter 15: Sleepy Head.

roam uncontrolled through memories. Of what is lunar dreamscape made but these? We'll soon explore the subconscious world of personal and typical (instinctive) mnemones, that is, major files of memories. Such world is not *per se* irrational but, carried passive on an incoherent dream-stream, an observer's various slumberous visions of it is. Sometimes, waking, he recalls his lucid travel or more jumbled narrative. There are even, in the loop, memories of dreams.

The Psychology of Deep Sleep

In deep, non-*REM* (or *NREM*) sleep the 'upper', voluntary structures of brain are cut from the loop. A sleeper's movements, including eye movement, are much restricted; sensation is dull or absent. Brainwaves, the overall coordinators of the central nervous system, slow to between 0.6 and 3 hertz. These are so-called delta waves. Maybe deltas drop to zero. Brain death. If, by head injury, stroke, tumour or poison, the sleep/ wake toggles fail or signals cannot reach the forebrain then the patient drops into oblivion. The curtain falls but drama does not start again. Coma is an open tomb, an unpinned shroud or wake-less sleep - though in its stillness deeper grooves of mind (archetypal constructions but also profound personal impressions and rote such as language) stay frozen yet intact.

Organisms sleep in different ways and yet, not dropping off into their 'little deaths' at all, they'd die. Sleep is vital. Survival is insistent, for example, that a brain is regularly cleaned (sleep's neural shrinkage lets in spinal fluid to wash toxins out). How, though, did genes evolve the physiology for 'off-line' maintenance or *REM* and *NREM* sleep? And, as well as chemical complexity, dormancy's a metaphysical affair; it knits up *mind's* (not brain's) ravelled sleeve of care. A just-so story for such wonder, best beloved, is that elevation from the 'psychology of physic' into coma must have woken life from its primordial lifelessness; and that, before you sleep on it, evolution really woke up when, atomically, the chemistry of slumber re-arranged itself as conscious beings such as you!

The Non-State of Consciousness -

The 'Psychology' of Physic

There's a rider tagged upon creation's tail. This state of subjectivity, if you could call it one, is its special case of total absence. Zero on the vertical, informative coordinate (see *fig.* 11.1).[89] You might call physics, chemistry and, bodywise, biology the 'psychology of non-consciousness'. *Their wholly automated, entirely physical behaviours are the only ones that naturalism can interrogate directly.*

---

[89] see Glossary: conscio-material (c-m) dipole and Chapter 11: The Basic Existential Dipole.

Frozen Time

You sleep but your past does not disappear. You wake and your past has not disappeared. You think you have forgotten, you may even suffer amnesia but untapped memories remain. They are how we freeze time. **A memory[90] is frozen time. It symbolically encodes the past.** *A memory is a thought object and, as such, has no life of its own.* A disc encodes music once recorded 'live'; it's a memory that, when replayed, affects the present and, from this, the future. Thus mental memory is an encoded image; it is a record and, on conscious recollection, becomes a presence of past action that may affect the future.

Bottom-up, mind is materialistically thought to be some aspect of dynamic brain and, from this neuroscientific but still philosophical angle, memory is a part of grey matter. Its storage bit, a hypothetical straw sometimes called an 'engram', is grasped as a 'nervous trace', 'a packet of proteins and lipids' or some other pattern of representation.

***Top-down*, neither memory nor knowledge are inherently physical.** No doubt, correlated nervous circuitry acts as a storage-and-retrieval system that, by association, allows the immaterial library of remembrance its efficient, selective interaction with an innervated body. Thus 'engrams' (are they sited at synapses, inside a nerve cell's body or elsewhere?) may, if they exist, indeed *relate* to physical experience; they may act as a recognition trigger, reference point or body's resonance with an experience. And organs (such as hippocampus and amygdala) certainly seem to log experience in the manner of a record/ playback head; they catch or release a moment that, in fixity, is called a memory. But if such 'storage' or 'playback' button fails the system's compromised. Either records are not made in the first place or the connection becomes impaired or irretrievable. *But the 'disc' of memory itself is metaphysical.*

On a cosmic as opposed to personal scale there is universal body (made of energy/ matter). If, equally, there is universal mind then there are archetypes. These act as informative potential; they serve as files lodged in the working program of creation. *It needs be re-emphasised that an archetype is a memory and, as such, a thought object. An archetype, as lifeless as a stone, has no life of its own.*

Memory, the only form of metaphysical information storage, is the shape of infra- or sub-consciousness. *Indeed, it is sub-consciousness.* **Sub-consciousness is made of memories.**

Regarding life-forms memory comprises an organism's library of precondition and conditions, that is, its context for experience. The precondition is its archetype, the basis of its sort. We might call this

---

[90] *SAS* Chapter 15: Frozen Time, Synchromesh 1 and Personal Mnemone; Chapter 16: Typical Mnemone, Archetype, Signal Translation, Instinct and Morphogene.

sort of memory '*typical*' while '*personal*' experience includes both active (created and transmitted) and passive (received) information. Such memories are not necessarily 'frozen' like a photograph. Nor, although it comprises a concept or basic expression of an idea, is an archetypal memory. Memories may operate like movies and store programs that, once triggered, can unfold like stories in a sequence to their completion. Any stored plan is, like a film or computer program, such a memory. Programs are, although dynamic, still a frozen form of mind; and they're replete with information. They specify the most efficient means to a well-defined end. **You might argue in this vein that biological structures are codification incarnate; and that the concept they express is an archetypal program.**

Psychosomatic Linkage

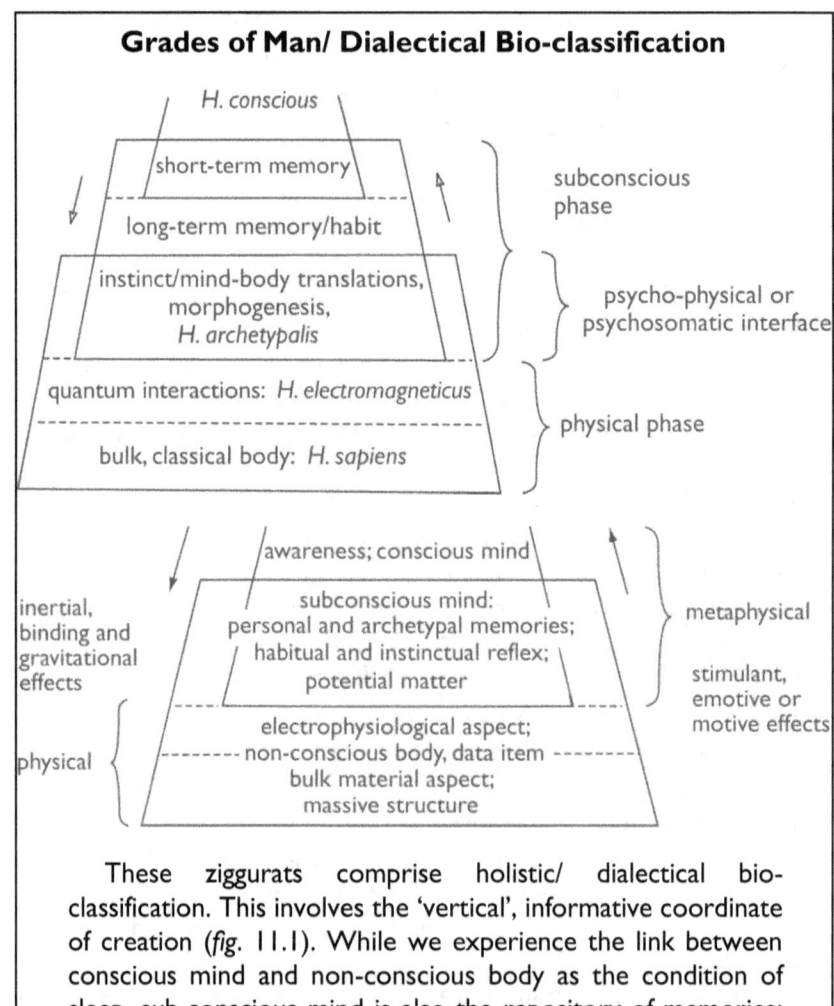

These ziggurats comprise holistic/ dialectical bio-classification. This involves the 'vertical', informative coordinate of creation (fig. 11.1). While we experience the link between conscious mind and non-conscious body as the condition of sleep, sub-conscious mind is also the repository of memories;

> *and these diagrams introduce a logical but revolutionary, top-down derivative from the conscio-material spectrum - the 'dormant' position of archetypal memory.* **Such channel of order is further identified with both the passive 'infrastructure of information' - codes including archetype - and with potential matter whence issue the various basic patterns of physics, chemistry and, with respect to complex 'bundles-of-laws' or 'programmed routines' called organisms, biology.**
>
> *fig. 13.5 (see also 13.1).*

For a materialist, who guesses informative consciousness is an 'outcrop' of brain chemistry, the next four chapters are irrelevant. Nor will you have met their content in the context of a modern scientific course. They are superfluous because, by materialistic paradigm, no dualism of separate mind and matter exists. Information and energy, mind and matter are of wholly the same substance. And if this substance is material it is non-conscious. Consciousness is therefore a peculiar effect of certain formulations of non-consciousness. How strange are mind and subjectivity!

Top-down, the psychosomatic sandwich (*fig.* 13.4) between conscious mind and physical body is practically inaccessible from either side. It is, in the upper case, subliminal and in the second, above whose body it is ranged, sublime. However, just because something is hard to access or measure scientifically does not mean it neither exists nor impacts our reality. *Natural Dialectic, noting the profound explanatory shortcomings of one-tiered materialism, simply and reasonably proposes an addition to physical electrochemistry - non-material, metaphysical mind.* What greater inherent improbability, wrote the founder of modern neurophysiology, Sir Charles Sherrington, than that our being should rest on two fundamental elements than on one alone?

In this case, what might constitute the nature of an interface?[91]

The basic principles of psychosomasis are clear. *Mind (at most gross, subconscious level) conjoins with matter (at subtlest, least-massive/ almost-immaterial level); elusive quantum probabilities pinned-down substantiate, it seems, certain processes; photon and electron precede, in the sense of underwrite, molecular and bulk reactivities; and, where electrodynamics describes the effect of moving electric charges and their interaction with electric and magnetic fields, biological electrodynamics precedes all bio-molecular considerations.* **Every biological process is electrical; and the flow of endogenous**

---

[91] *SAS* Chapter 16: *H. electromagneticus*, Psychosomasis and How Does the Connection Work? Also Chapter 17: The Logic of Embodiment and *fig.* 15.5.

currents is the primary and not secondary feature of physical life. Not only biochemistry but quantum biochemistry heave to the fore. Natural Dialectic lifts perspective from molecular to a vibratory, field perspective. It is thus suggested that, at electrical and wireless levels, patterns of subconscious mind meet and influence matter; archetypal information is relayed to chemistry by polar charge and light.

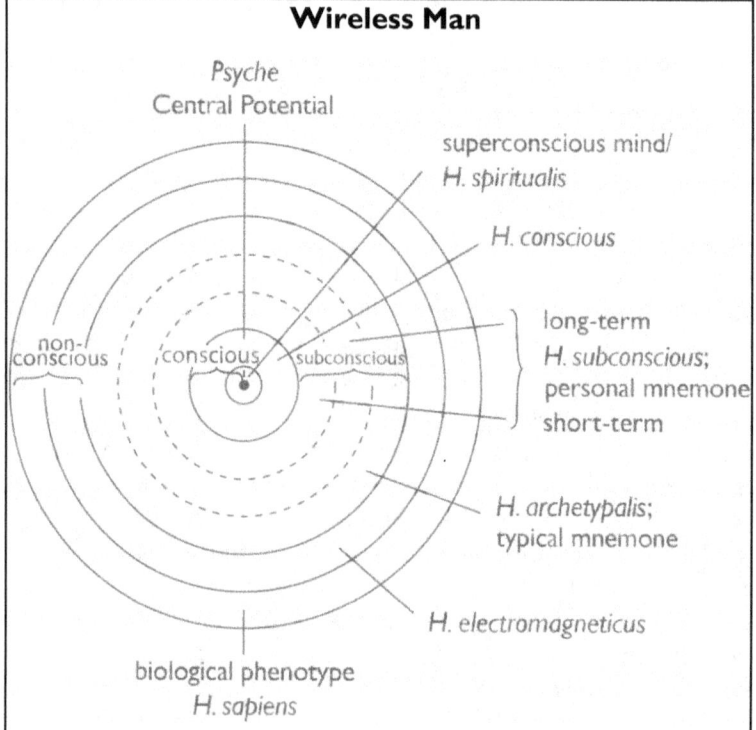

In this spectral representation of human grades, sheaths or bodies only *H. sapiens* is wired, fixed or made of classical matter. The others are wireless or radiant.

Thus you might visualise yourself (and any other organism) as both wireless and a wired anatomy, that is, as a composite of fixed and fluid sheaths. As discrete electron orbitals surround a nucleus these sheaths surround the Nuclear Psyche. Separated by 'an exclusion principle' their phased 'energies' materialise towards the periphery of creation. There the most dense, wired level - called your biological body - is obtained.

fig. 13.6

Such consideration leads to the picture of a human mostly wireless! Note that, in this human case, *H. archetypalis* is composed of an archetypal memory called his typical mnemone.

The Typical Mnemone

Materialism ridicules the idea of natural informative structure, a metaphysical 'cloud' or mnemonic database called archetype. Where, however, does its absence leave us? **It is mind not molecule anticipates and innovates.** *Might immaterial elements conspire with the material components of biology; might rational mind conceive the mechanisms that can serve its purposes; and subconscious control (the exercise of a mnemonic archetype) inform all forms of life?*[92]

No atom or group of atoms struggles to survive, innovate or build integrated systems. **Thus, as regards all life, is not the fundamental error of naturalistic methodology to deny any element or force dissociated from matter?** Perhaps the indulgence that it's 'all in genes' or 'simply chemical'[93] is a professional deformation and, to boot, a philosophical delusion. If you consider information, reason and mind's other faculties to be material-only entities when, in fact despite consideration, they are not - what of materialism then? What follows makes sense, in this light, of your internal metaphysic.

It bears emphasis that subconscious memory is uncreative. Its record is, of itself, passive, purposeless and fixed. *And yet its reflex is, like that of any technology, originally the product of creative purpose; and such purpose is the motive behind the otherwise motiveless government of biological bodies.* **It is the origin of meaning and rationale in biology.** Let's take a peep, therefore, at the distinct rationale behind the psychosomatic nature of *H. sapiens* - you. It has already been identified (*figs.* 13.5 and 13.6) as a typical mnemone called *H. archetypalis*.

H. archetypalis, the Image of Man

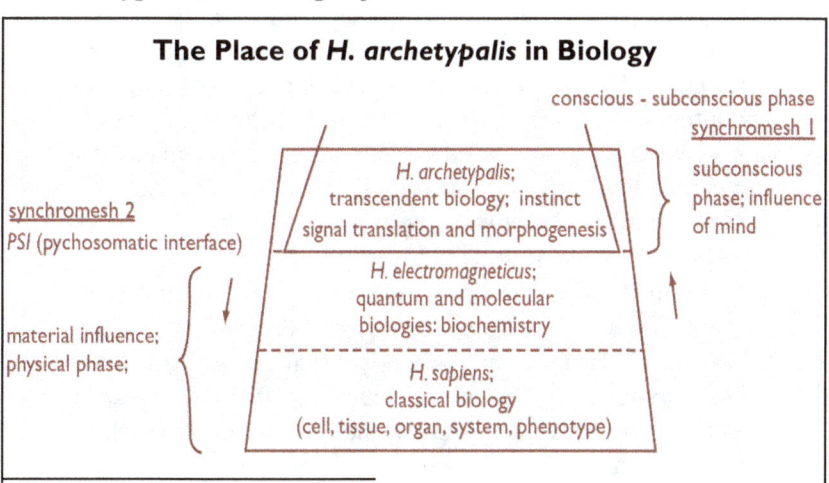

---

[92] Chapter 15; also *SAS* Chapters 6: Information's Infrastructure - Code; also 16 and 19 *passim*.
[93] see *A&E* Chapter 8: A Book Needs and Index esp. the first link.

Check *fig.* 13.6. Metaphysical archetypes such as *H. archetypalis*, 'the image of man' or 'memory man', are *not* (because they involve the immaterial, informative coordinate) the same as any taxonomic definition of species. In the dialectical view information is broadcast from archetypal level to its biological correlate. Such local receivers are cells or multicellular bodies. Generic program acts as a homeostatic regulator, a reference-point or 'norm' that integrates purposive developmental sequences, chemical algorithms (such as metabolic pathways or mitosis) and the morphogenetic disposition of pre-coded building supplies for growth, maintenance and repair. An archetype is, in effect, a lawful program. Each one represents a statute (the organism as a whole) with its bundle of clauses (sub-routines). Informative principles or laws derive from a level above. The level above matter is mind - in this case a sub-conscious store of logic called an archetypal memory. As your own behavioural patterns derive from a context of memories, so do those of nature. Indeed, are not your instincts and their body part and parcel of a very natural pattern, the physical expression of a frame of reference, a metaphysical criterion called **deep and natural memory?**

Mind naturally encodes information. Subconscious mind consists of coded files that we call memories. It is suggested that passive, archetypal files inform, at the lowest level of existence, material energy. Every cell partakes, as an antenna, of archetypal as well as electromagnetic guidance. This couple acts as a regime within which correlated sub-routines correspond to, entrain and control the form and function of physical counterparts. Their program, in conjunction with 'read-only' material script called *DNA*, generates and maintains physical form; but why, any more than a TV program's broadcast code resembles what it generates on screen, should archetype resemble bio-form?

If you think 'archetype' is just a 'cop-out' in explaining how things are then I suppose a systems analyst must think conceptual plans for any working mechanism are 'cop-outs' too. However, philosophical objection to a program's purpose in no way mitigates the impact of its natural possibility and, if an immaterial element of information exists, natural fact.

fig. 13.7.

Fig. 13.7 indicates that a typical mnemone is composed of three main sub-routines; or, if a protocol is a standard procedure for regulating the transmission of data between two end-points, three linked protocols.[94] **Together translational, instinctive and morphogenetic programs comprise an individual's archetype.**[95] **If, as in the case of plants, fungi etc., there is no conscious component then the translator element is reduced from nervous to chemical (e.g. hormonal) messaging alone.** *H. electromagneticus* **has, in the human case, been identified as the physical side of message-exchange and** *H. archetypalis***, with its routines, the metaphysical correlate.**

Two important aspects of archetype are marked. **The *first* is that, just as the genetic 'book of life' is found as a nuclear genome in every cell, so every cell accompanies its typical mnemone. The *second* is that such generic 'broadcast' survives the death of any individual cell or body.** Genetic material is passed from generation to generation with its accompaniment, a cell; the physical medium of transmission is, of course, a single fertilized egg or product of asexual division. The metaphysical mnemone is not thus passed; you might liken it to a generic permanence of which any ephemeral cell is the attuned receiver; destruction of local apparatus does not affect the broadcast. Nor can archetype become extinct.

In this view typical mnemone is the metaphysical blueprint for a each type of body; it is body-linked at the level of a cell. Cell not soul has biological memory. Each one also contains genetic code whose switching systems produce appropriate materials for its own sustenance and, where applicable, communication with the rest of the multicellular operation called a body. In the case of a hologram different pixels appear differently, suppress or lock out parts of the whole picture but still contain its whole potential; similarly, in the mnemonic case, 'a pixel of the archetypal hologram' interacts to excite or suppress an appropriate cellular routine from the whole program of its organism's life-cycle. Possible mechanisms of mnemonic/ genetic interaction are suggested later in this chapter.

Thus a cell or body's typical mnemone is seen as a morphogenetic attractor, an 'intelligence' that moulds substance and thereby coordinates the production of functional shape - shape whose function relates specifically and accurately to its purpose. **That is to say, all cells partake of memory.** *DNA***'s material book of life is correlated with a book of immaterial information in the form of archetype. Each cell responds, as well as to its circumstance, to its subconscious archetype.** As it holds genetic potential for all chemicals the body will employ, so

---

[94] for more detail see *SAS fig.* 15.5.
[95] *SAS* Chapter 16.

morphogene holds files for every shape; in this metaphysical sense each cell includes its adult's form!

For example, human cells each resonate with *H. archetypalis*[96] and each cat cell with *Felis archetypalis*. So, within the whole picture, does each tissue or organ made of cells. Each pixel partakes of and corresponds to the whole. Just as a cell can express only its own relevant fraction of the genome, so it resonates with its relevant fraction of archetype. Individual sub-routines or linked groups of them are accessed within the framework of a coherent whole. In this way the archetype is a complete context, a governing template whose different sub-routines are accessed as parts of a coherent master program. Thus body cells and unicellular organisms alike 'know' their behavioural patterns. **This knowledge is not conscious; it is unwitting as sub-conscious means and is.** *But all cells have dormant mind that interacts with their chemistry, chemistry that includes physical correlates of their 'dormant intelligence' in the form of information systems (cell signalling, hormones, nerves etc.) and a database of DNA-written coils (or solenoids?).* Nor is this passive form of mind less separate from atomic action than a dye from its cloth or a solute from solvent. It is closer, close as broadcast is to the picture on its TV screen. Closer than a nanometre, mind substantiates material expression. It is, although materialism misses its dimension, a control centre.

Finally, *figs*. 13.6 and 13.7 also indicate that, entrained with metaphysical archetype, the physical side of the psychosomatic interface is composed of an entity dubbed *H. electromagneticus*.[97] What is this?

H. electromagneticus

A metaphysical point of reference, H. archetypalis, is physically reflected or, if you like, realised by its electromagnetic correlate dubbed, dialectically, H. electromagneticus.

For physics quantum fields are the basic ingredient of cosmos; particles are simply specific bundles of energy and momentum in their field. In this sense *H. electromagneticus*, composed of charge and light, is a quantum field representing the the phase of quantum biology. No doubt, as active sites in enzymes are framed in 3-D space by molecular arrangements, *H. electromagneticus* is framed by molecular and cellular shapes. This framework allows specific, highly precise quantum interactions with its metaphysical archetype, *H. archetypalis*. Physical events are thus governed by a metaphysical component itself free of physical constraint, locality or temporality.

Thus, not surprisingly, in 1935 Dr. H. S. Burr, Professor of Neuroanatomy at Yale University, and Dr. F. S. C. Northrop established

---

[96] *SAS* Chapter 17: *H. Archetypalis*.
[97] *SAS* Chapter 16: *H. electromagneticus*.

that all 'living matter' from a slime-mould to an elephant, from a seed to a human being, is surrounded and controlled by electrical fields. Indeed, every cell pumps ions to maintain a healthy, electrical potential across its lipid membranes.

Electromagnetism involves photons. Is light from interacting electrons not in fact the language of life's cells? Fritz Popp, founder of the International Institute of Biophysics at Neuss, Germany, proposes that low-level light emissions are a common property of all cells. Such weak luminescence ranges from thermal (infra-red) to ultra-violet. Not only do electrochemical forces across cell membranes help control their permeability. Colleague Marc Bischof believes that weak, coherent e-m fields combine to regulate not only the cell's surface but its internal members. Thus, correlated with the positions, densities and movements of electrons, a 'signal web of light' might harmonise cooperation of organelles with each other and with chemicals throughout the cell. Moreover cellular cytoskeleton's coiled, semi-conducting filaments and tubules, whose other jobs include structural support and transport track-ways, compose a network for the conduction of charge and generation of electromagnetic fields. With this, as well as body heat, living organisms faintly glow.

These 'wires' of electrodynamic propagation may power up other structures such as protein alpha-helices and coiled/ solenoidal *DNA*. Indeed, Bischof postulates that *DNA* pulses as a 'light pump', that is, as if an aerial both emitting and absorbing light. As brain-waves control the *CNS* such a mobile web of light would, with its various frequencies, control cellular operations. Cannot radio carrier waves incorporate signals we tune into and call programs? Similarly, might light signals control not only internal operations but also employ cytoskeletal components, where they attach to the surface membrane, to form electrical and possibly 'fibre optic' channels to the exterior matrix? This extra-cellular matrix (*ECM*) provides a medium for body-wide bioelectrical linkage. It is, therefore, possible that Burr's L- (for life) fields coincide with the description of cells as electromagnetic units and so with the idea of a mass-free lattice identified (*fig.* 13.7) as a quantum-level sheath called, for organisms in general, *X. electromagneticus.*

To summarise: radiant systems, whose weightless quanta represent the purest form of physical energy, are logically identified as a prime candidate for the physical side of the psychosomatic exchange of information. It is therefore suggested that psychosomatic exchange occurs between a typical mnemone (with its three parts of signal translation, instinct and morphogene) and molecular structures through the medium of highly organised electrical, electromagnetic and luminous vibrational patterns. These patterns are, as the active site of an enzyme is framed by the conformation of its protein chain, framed by gross matter; but, inwardly and equally, by archetypal program.

Thus the interaction of mind with matter is seen as a dynamic, coherent association of two anatomies, wired and wireless. As quantum world rules gross appearances so wireless orchestrates the wired. The 'wires' are bio-molecules and other matrices, such as nerves, with electronically active sites. In this case fixed psychological harmonics called memories interact with body through the medium of photon, electron and atomic bond vibrations. In such association atoms and molecules of your visible body, *H. sapiens*, act as oscillating resonators. *Thus interlocked, Janus-like psychosomatic bodies can be seen as sub-conscious information, in the form of memory, acting in concert with complex electrical fields. Sub-conscious mind is the hand, electricity the glove.* **There is nothing in biology to suggest such concert does not exist, indeed there is much to support it.**

Psychosomasis

Psychosomasis is the way psychological meshes with physical 'below'. It concerns, in this case, the junction of all three aspects of typical mnemone with the physical side of the psychosomatic interface - an electrodynamic, resonant medium called variously your 'network of light' or, aforementioned, *H. electromagneticus*. Psychosomasis is simply a word that identifies the process of translation, a transduction of information between *H. archetypalis* and *H. electromagneticus*.[98]

There is no doubt, the universe is in vibration. The cosmic transmission of information and energy is, at root, vibratory. Ordered oscillation is called harmony; harmony is the grammar of music and music is a universal language. The theory of music is implicit in any recital. Could it transpire that explicit order of a cosmic recital is the product of implicit notation? Could its excitement represent cooperative forces, specific 'notes' called particles and thereby, all in all, harmonic code?[99]

Why, on this chime, is the supreme cosmic model harmonic vibration? **Why is the Master Analogy of Natural Dialectic music?**[100] Cymatics (the study of the effect of waves on matter) has found striking evidence of the patterns that sound can induce on large-scale, let alone quantum-sized, matter. For example, by 1787 Ernst Chladni had drawn a violin bow against a thin metal plate covered with particles of dry sand and reported the resonant effect of given frequencies of standing wave. He watched, in a direct and obvious example of morphogenesis, implicit energy inform explicit shape.

---

[98] Glossary; also *SAS* Chapters 16 and 17 and Index: psychosomasis and polarity psycho-biological.
[99] see Chapter 11: Passive Information and Chapter 12: Information's Infrastructure - Code; also *SAS* Index: alphabet, cosmo-logical language.
[100] *SAS* Chapter 6: Music.

More recently Swiss Hans Jenny invented a tonoscope which, using crystal oscillators to precisely vibrate plates or membranes, converts vibrations to patterns. Thus sounds uttered into a microphone yield visual representations on a screen. When vowels of ancient Hebrew and Sanskrit (but not other languages) were pronounced the vibrated particles took the shape of their written symbols; and when the Hindu sacred syllable for primordial creation, *Om*, is correctly intoned it produces a circle (representative of the Infinite Void from which all things issue). The diagram called a *yantra mandala* symbolizes an entire, sound-drummed cosmos. It is a perfectly symmetrical picture radiated from the Sound of Silence, from the Nature of Nothing. **Such harmonic resonance is foundational.** It raises strong possibilities, always avowed by the mystic, that sounds and names have internal properties of their own. Such 'intonations' are a fascinating reminder of, at physical level, the connections between sound, geometry and the development of form.

Resonance, whose orderly aspect is characterised by an analogy with music, is the tendency of a body or system to oscillate with larger amplitude when disturbed by the same frequencies as its own natural ones. *It therefore intimately involves the vibratory transfer of energy.* Such transfer is an integral part of all vibratory systems wherein waves interfere with/ destroy or cohere/ amplify each other. Science is familiar, for example, with nuclear magnetic, electromagnetic, electron spin, acoustic and mechanical forms of resonance. Acoustic (musical) resonance involves the sympathetic vibration of stretched strings or air in pipes; thus it is easily understood that, as well as tuned circuits, quartz crystals and so on, musical instruments are resonators. Indeed, the motion of waveforms is closely associated with harmonics. Sound waves are governed by the law of harmonic relationship whereby notes of the right frequency combine into chords, and chords and notes in time into harmonies. Resonance occurs when one note or chord vibrates in harmony with components of notes or chords in a different octave. One body vibrates in sympathy with another. A similar sympathy can occur in the oscillations of electrical and mechanical phenomena. Energy is amplified and transferred by resonance and attunement. Common examples of electromagnetic resonance include tuning a transmitter/ receiver and photo-electric initiation of the photosynthetic process, that is, the first step in life's chemistry.

For quantum physics matter is certain vibratory frequencies of energy; and, from a dialectical point of view, it is simply stresses, strains or tensions in the medium of their absence, that is, nothingness. **There is, however, nothing random in a highly orderly creation derived from first acoustic principles.** The universe is, physic agrees, a kind of machine finely-tuned according to about twenty fundamental physical constants, archetypes or notes. Could these notes constitute the instruments that shape such chords as elicit an electron, proton, atom, molecule and thereby, vastly larger and more orchestrated, you? *Like organ pipes such instrumental archetypes would be the shapes of resonance through which the notes of cosmos can appear.* **The idea of musical archetypes as a source of physical order may alienate one-tiered materialism but its synergy dances to the tune of Natural Dialectic's Central Metaphor.**

What is being proposed is, in other words, vibration prescribed by standing waves. Such information is, essentially, melodic. The resonance patterns that a fundamental string can support would give rise to the properties of an elementary particle such as an electron (*cf.* the harmonic equation for standing and probability waves). Motion, energy, mass, force, charge, and so on are determined, in effect, by the precise, vibrant events that a particle's internal strings execute in symphony. This 'unstruck music', a particular 'chime' or 'intonation', is its hallmark. Just as a spectroscopic analyst understands that each sort of molecule emits its own 'fingerprint', 'signature' or 'bar-coded *PIN*' by which it is uniquely recognised, so different 'chimes' show as the forces and particles of nature. Waves both initiate and lay their fingerprints on form; and, as with musical theory, the rules authorise permutations like chords rather than discords - a system fundamentally harmonic and not cacophonic.

Are atoms harmonic oscillators? Are quantum harmonics reflected in larger visible dimensions? Are crystals an example of this coordination? The answer in each case is yes; and the effects of harmonic vibration on, for example, jets of gas and flame in air are dramatic. You can also make 'electromagnetic plants' whose flowery patterns arise in a dish of ferrofluid placed over a single iron-core AC coil; the fluid accords with lines of the field. And it is significant that Chladni's figures often imitate familiar organic patterns that we see in nature and, especially, biological structures. Pattern clearly relates to frequency of cycle. The ancient architecture of leaf, fern, wood grain and many other kinds of morph appear. Music is frozen. Orphic sound and archetypal, Pythagorean geometry are thus combined.

In this view inner action naturally precedes and governs outer structure. Visible expression of energy is the inverse of vibratory pattern; visible mirrors invisible. *Between the 'dead' ground of material shapes course channels of energy - the invisible rules.* **Oscillation between polarities, cycles, vibratory rhythms, the interrelationship of waveforms and complementary resonance are at the heart of Natural Dialectic.**

How Does the Connection Work?

What, therefore, is the psychosomatic grail? What earths mind to matter? How, mind > body/ body > mind, does the connective 'ligand' work?

Rhythmic beat, coherence, harmony - their influence draws the whole world into order. It chimes bells and dances. Music vibrates shapes and its songs, each in their own way, feel right. Thus natural music of our cosmos is expressed. *The heart of morphogenesis is buried deep in resonance.* **A key phrase in the suggested explanation for the wireless, psychosomatic transfer of information between subconscious structures of the mind and the physical plane is 'resonant association'.**

Internet, television, wireless - energetic frequencies can carry information; antennae are tuned for resonant association with specific frequencies; and, since the emission/ absorption of electromagnetic radiation by atoms and molecules peaks at each one's natural frequency, they act as aerials. **It is thus suggested that psychosomasis between mind and body is, with appropriate devices in place, common and continual.** On the *mind's side* of the interface the device is called a mnemone; on the *body's* it occurs when the appropriate molecular/ cellular bioelectrical configurations are in place. Synergetic resonance occurs whenever thought affects nervous/ muscular activity or sense data is transferred for symbolic appreciation in mind. The logical postulate for such informative transfer of harmonic codes is the most subtle, least physical substance - photons.

An argument was developed suggesting that the principle of psychosomasis[101] was, mind <> body, resonant translation at an electrodynamic/ quantum level. **In this respect, regarding the relationship of nuclear archetype with electromagnetic configuration, resonance sources communication; and its agent, matter-in-principle, is vibrant, quantum forms.**

The modus operandi of psychosomatic broadcast is therefore, in a word, attunement. Resonance. It involves transduction between recorded information (memory) and physical energy.

In short, mind is linked to matter by a wireless anatomy whose instrument is resonant association.

Whence did the Wired Side Emerge?

If the 'psycho' side of psychosomasis is wireless its somatic side is wired. The body's wired connections are, as previously suggested, bio-molecules and other matrices, such as nerves, with electronically active sites. Whether it is, or develops from, a single cell, an earth body is a fully codified, integrated 'wired' system.

Do we therefore suppose a cell's or body's perfect health evolved through grades of imperfection, that is, millions of sicker stages? Or prefer that archetypal vigour radiates original, dynamically fine-tuned

---

[101] *SAS* Chapter 8: Principles of a Unified Theory of Matter; Chapter 15: Psychosomatic Linkage; Chapter 16: Psychosomasis.

health for every kind? *If your body's health is indeed the result of interaction between physical and archetypal (sub-conscious) patterns, it did not evolve by accident.* The tendency of every bio-system is to bounce back to its archetypal norm, its *Vis Medicatrix*. **This metaphysical form is the very basis of homeostasis, physical life's overriding principle. Health is the norm.** Wounds heal, infections are fought and cell debris cleared. Indeed, the central nervous and autonomic, endocrine and immune systems are integrated in such a holistic way that experts sometimes use the phrase 'psycho-neuro-immunological system' to describe their cohesion. Mind, as every doctor knows, affects matter and *vice versa. The channels of 'psycho-somatic effect' are precisely what this chapter is describing - unless you still neuro-scientifically believe that mind is brain and psychological equates, essentially, with chemical.*

Swing life round to the other side. Instead of reason seek, as rational Darwinians do, its absence. Is, as Lady Luck avers, the irrational cause of reason chance? So coded neuron chemistry evolves and wiring programmed brains needs no intelligence? Yet reason floods biology. Metabolic pathways, cells and code-specific systems all express, in different organisms, common physiological functions. These functions represent, effectively, 'reasons' subordinate to the overall 'reason' that drives a biological body. This 'reason' is to live more life, that is, survival. Homeostasis buffers changes in and outside bodies. Such equilibration promotes balanced working, that is, survival. 'Sub-reasons' therefore include sensitivity, biochemical command and control, digestion, respiration, osmoregulation, reproduction etc. *Which of these functions is, where found, superfluous?* Which of its associated organs, metabolic factors or coded molecules is, lacking a reason for existence, therefore redundant? *Which can survive without the others?* **Calculate the minimum number of functions necessary to support life, any life at all.**

Reason's nowhere absent but, with chance, like chalk and cheese. Reason is meaningful, chance meaningless. *Why should a highly rational system irrationally 'self-organise'? Chapter 15 will demonstrate the fallacy of a belief that any chance whatsoever could invade the initial, entirely purposeful construction of a cell, a human or any other type of biological body.*[102] *The chemical evolution of proto-life is a fantasy.*[103] *And your own human microcosm underlines the utter feebleness of evolutionary explanation.*[104] If materialism's rational it spotlights how irrational, backing chance as its creator, rationality's become.

Isn't it, on the holistic hand, inconceivable that such a logical,

---

[102] *SAS* Chapters 19-25.
[103] *A&E* Chapter 9; *SAS* Chapters 20 and 21.
[104] *AMA? passim*; *A&E* Chapters 6 and 17; *SAS* Chapter 17.

integrated, self-consistent embodiment as yours, constructed with highly specific complexity in accordance with grades of the conscio-material gradient of creation itself, occurred by chance? **If reason wins, chance and time's grand theory crumble back to their home ground - they bite their progenitor, the dust.**

Triplex Psychology Summarised

In summary, *bottom-up* materialism theoretically reduces consciousness along with both conscious and subconscious forms of mind from metaphysical to physical status. The presence of metaphysical (psychological) parts is denied.

From a *top-down* perspective there is, however, no such denial. The metaphysical/ immaterial factor is (see Chapter 11) information. And from the hierarchical perspective of holistic Natural Dialectic the subconscious phase of mind is sandwiched, as computer software, between conscious programmer and physical, bodily hardware. For the holism of Natural Dialectic it comprises two integrated parts - metaphysical archetype (*H. archetypalis*) and, at the psychosomatic interface (*PSI*), physical *H. electromagneticus*. The subconscious amounts to an inner, psychological context composed of memories both personal (uniquely individual) and typical of a kind of organism. The latter, typical mnemone is, in turn, composed instinct, morphogene and, where consciousness is involved, signal translation. Two of these inhabit, like *DNA*, every cell in every body. The third, signal translation, occurs in nervously endowed animals.

In the latter case two-way flows of inward sensory and outward motor information are translated from nervous into conscious, experiential forms and *vice versa*. These anti-parallel, in-out vectors characterise the whole information system. In the nervous case both synchromesh 1 (the conscious/ subconscious 'gear change': see *fig.* 13.7 and *SAS figs*. 15.5 and 15.6) and also synchromesh 2 (the phase change at *PSI* or the psychosomatic interface) are operative. Incoming sense messages impinge, via signal translation, instinct and the 'library' of personal memory, on conscious mind; and, in the outward, motor direction, either a purpose or instinctive reflex instructs body by way of signal translation, instinct and and morphogene.

In dormant, nerveless organisms, on the other hand, only synchromesh 2 is operative. Incoming messages from the physical side impinge on morphogene and instinct; and response, if any, tends the other way.

In ourselves the whole system amounts to an inner, psychological context which, in conjunction with a biochemical 'chip' called *H. electromagneticus*, autopilots mind and body (including brain); its bio-cybernetic circuits keep the plane flying while its conscious pilot concentrates on his chosen, narrow screens of data. In other words, it constitutes an exquisite, involuntary operating system backing up a voluntary operator.

There is no doubt the whole business is highly, exactly programmed. Its metaphysical and physical components inhabit, like *DNA*, every cell of each body. That is to assert, mind with electromagnetic fields inhabits *every* organism even if many (i.e. plants and bacteria) are wholly dormant. *H. electromagneticus* involves electromagnetic fields with their electrons and photons. If you could strip cell structural material, tissue, organ and a whole body of all atomic and therefore molecular and bulk mass, the residue would constitute the quantum biology of signalling and connectivity.

As in physics, the field is primary. Its ubiquitous bio-electrical activity informs and controls secondary, passive structures. Of the latter the cell 'brain' (or membrane) with its two-way, in-out flows of message and response is one; the behaviour of intracellular biological machinery in the form of changing protein shapes, the specific motion and interaction of biochemicals, the operation of organelles and the dynamic data bank of chromatin (that is, genetic blueprint and epigenetic signalling) also act in response to either external or internal prompts. In short, the whole homeostatic process of living bodies is instructed by the field(s) we label *H. electromagneticus*.

As regards conscious mind-to-body communication the body's overall controller, brain,[105] appears to concentrate psychosomatic transfer between *H. archetypalis* and *H. electromagneticus* at its centre, called the 'interbrain' or limbic system. This system exchanges information between the forebrain (or prefrontal cortex, the immediate 'wire' onto which a pilot's thought messages are, like morse, tapped out) and midbrain, hindbrain, brainstem, spine, peripheral nerves, glands and immune networks. It involves organs associated with emotion (the amygdala) and memory (the hippocampus); its diencephalon includes a crucial triad of glands - pineal, pituitary and hypothalamic - which link electrical (nervous) and chemical (hormonal) systems of information carriage.

Finally, chance and rudderless necessity could not originate the wired side of life. Bearing this in mind we are now in a position to turn from metaphysical mind and, at the physical side of the psychosomatic border, *X. electromagneticus* to, firstly, the energetic patterns of physics and then gross, codified bodies of biology.

---

[105] *see* also *SAS* Chapter 13; *SPFP* Lecture 3; *RSP* Chapter 3.

Chapter 14: Triplex Physics

The world's a stage. Physics describes our platform and its props. Let's start by seeing how it might have started.

Did the universe create itself? Nothing can create itself.

Is energy eternal? If so, what of entropy? By heat death the world would, infinitely long ago, have become inhabitable.

Cosmos comprises the entirety of physics. Thus its start from nothing physical must have been metaphysical. All matter issued from an immaterial source.

In this case the three levels of physics externalise from an internal source. The 'top', metaphysical level is potential energy/ matter; there follow quantum and 'locked', classical phases (which latter we appreciate as bulk matter). **They compose a triplex scale of objectivity.**[106]

(Sat) Potential Energy

Potential precedes possible action. It is a prerequisite or precondition for results; but Natural Dialectic does not think of physical potential in the way that physics does. Potential matter is, as we've seen, a metaphysical, informative affair.[107]

Informative archetype (potential energy/ matter) is thought of as letters in a universal alphabet, as bits and bytes of a computer program or, better (since vibrations/ wavelengths correlate with forces and energies), notes whose harmonics compose the cosmic opera.[108]

Precondition

***Bottom-up*, no metaphysic orders what is physical.** The so-called 'laws of nature' are but descriptions of behaviours that a starting point (say, big bang/ initial projection) has spawned. Formulations and the values of their constant factors are not fixed by any deeper level than experimental regularities. Natural reasons have no reason; nor do material effects have *a priori* cause. Certainly there's no intention that substantiates cosmology; the character of automatic movements started as an accident; freeze teleology right out of every frame - this is the order of the game!

---

[106] Chapter 11: Lower Pole; Energy; Chapter 12: Upper and Lower Syntactic Levels.
[107] see *figs.* 7.3 and 8.2; also Chapters 9-11; see also Glossary: archetype.
[108] see *SAS* Chapters 6: Music; 11: Nature's Holy Ghost; 16: Psychosomasis and How Does the Connection Work?; also Index: music, archetype, mnemone and harmonic oscillation.

Invariance under transformation. Underneath the world's commotion basic character of parts remains the same. Contexts change but automated rules of play do not. Without conserved invariance you can't obtain the balance that equations need. Energy's conserved; at any time in any space from any angle laws of physics stay the same. **Does physics see such changeless cosmos as a self-consistent, fine-tuned set of principles? Or are its invariant patterns of behaviour basically informed by chance?**

Top-down, **the order of invariance is hierarchical.** Cosmos, not chaos, constitutes reality. A stack expressing law might read:

	tam/ raj	*Sat*
	existence	*Essence*
	exhausted/ kinetic phases	*Potential Phase*
	variation	*Invariance/Permanence*
	development	*Seed*
	practice/ actuality	*Principle/ Law*
	subsequent order	*Archetype*
	elaboration	*Plan*
	change/ inconstancy	*Stability/ Constancy*
↓	*tam*	*raj* ↑
	passive state/ object	*active energy/ force*
	exhausted phase	*kinetic phase*
	informed/ ruled	*informant/ ruler*
	contingent variation	*basic themes*
	external 'fall-out'/ outcome	*internal order*
	crystallisation/ bulk shape	*energetic pattern*
	chance	*necessity*
	accident	*design*
	apparent chaos	*cosmos*

Natural Dialectic indicates a source from which the stages of a cascade fall; it involves an inner, central 'symmetry' from whose intrinsic order cosmos breaks. It breaks, like all phase changes, with a natural spontaneity. Natural, contextual spontaneity, however, never barred preordination. *Rules are information, information is potential for behavioural patterns.* Regulation thus precedes and guides the way a game is played spontaneously. Solids fall from gases and, as they do, so different rules of state apply; different sets of laws precipitate from inwards outwardly. There would be, according to this scheme, essential principles and primary law; and secondary sub-routines that follow on. What is, we'll need to keep on asking, the central source of symmetry, polarity and bifurcating cosmic code?

Law is regularity; symmetry of law is stable; and, as Einstein claimed, the rules of nature are 'incorporated reason'. Archetypes, in this encoded sense, are the voice of reason heard through mechanisms of material form. Their issue is a 'breakage' out of background metaphysic; expression leads from hidden symmetry through simple, subtle quantum agents to the detailed aggregates of flux and sharper fixity. As scientists know and strive to better understand, you have to dig to find the principle substantiating practice. *From what superficially seems chance you have to disentangle rule.*

Lack of Cosmo-logic

Physics is defined by formulaic rules and yet, at root, materialists conceive of Chance as Grand Creator. Is chance reasonable or, by definition, not? Yet Lady Luck, it's claimed, gave birth to cosmos and to life. Such Reason isn't reasonable but, still, this face of Mother Nature is a major player in the scientific pantheon. Unfathomable Fortuna needs attention.[109]

Do chance, uncertainty and probability reside not so much in large-scale transformations as in the very substance of material constructions? A change is labelled 'unexpected' not because of nature's operation but because a relatively ignorant observer calls it so. **If, however, large-scale things are made of quanta what's the basic nature of reality?** A 'cloud of metaphysic' (probability) collapses into actuality; which of this pair, determinate or indeterminate, best represents the ultimate material verity? If, moreover, indeterminate substantiates determinate how do they correspond? How and why should a 'collapse' of quantum probability perpetually render large-scale certainty? *How, in short, do relations between classical and quantum mechanics work?*

If, further, 'quantum strangeness' is the norm, why is regularity apparent everywhere? Mathematician Roger Penrose takes an inclusive, common-sensible approach. From a quantum perspective 'bizarreness' *is* normality. It exists in potential but is 'confined'; it's everywhere a possibility but is, for the most part naturally 'suppressed'. How? To observe a quantum you disturb its state; you collapse its 'probabilities' into an actuality. Local fact is realised. You don't, however, need a laser or some other kind of probe. Continual, interactive impact by bulk objects of the universe engenders mutual interference and, collapsing probability, forever brings the action 'down to earth'. Past and future tryst in present actuality. An immaterial 'cloud of possibility' is realised; immanent potential rains with drops of matter locally expressed. Can't quantum 'fluff' and classical 'hard angles' co-exist? Constrained by billions of neighbours most quanta do not dance in isolation; bound in molecular and larger entities they submit to various chemically and physically predefined forms of order called the certain, non-random laws of nature. Creation *is* the orderly constraint of energetic chaos.

Possibility to actuality; non-location to location; what translates every-where-in-principle to here-in-practice? Check the previous stack. *Quantum*

---

[109] *SAS* Chapter 7: Getting to Grips with Lady Luck Shots in the Dark etc.

uncertainty seems to reflect creation's field; for Natural Dialectic it's a cosmic shadow whence our 'real' world appears. **It *partially* describes the staged translation of potential matter to expression - wave agency and then, with character intact, collapse to property-of-particle whence every atom, molecule and other shape is locally and orderly constrained.**

Thus, if all systems are at root 'non-local' then physical nature appears supported by an invisible reality whose connectivity is omnipresent. One might thus accept that the 'Copenhagen' way of thinking - based on Heisenberg's Uncertainty principle - was a true description of 'microscopic deep reality'.

On the other hand, you might accommodate determinism. Perhaps, as Max Born wrote, 'the motion of particles follows probability laws but probability itself propagates according to the law of causality.' David Bohm's account of quantum theory is also deterministic. And 'God does not play dice', said Einstein. No doubt, in any case, we're here because the universe is as it is. That's no surprise. How came it so? Was its initial condition chance or not? The universal body is sharply defined by a precise set of over thirty interdependent settings. *Their values combine to generate a universal pin-code that was either preordained or at least intrinsic in primordial projection. Indeed, the dials are set for the sun, earth, you and me to an accuracy computed by Oxford mathematician Roger Penrose at 10^{10} to power 123!* **If true, that cuts chance completely out.** Erasure of coincidence. The probability of your bullet hitting a nail-head at the other end of cosmos first shot is vastly greater. Mathematicians consider odds longer than 'only' 10^{50} against to be zero. That is to say, there is statistically no chance whatsoever that cosmos and its dependent life are accidental. **The Penrose computation, if valid, indicates that odds against the observed, law-abiding universe appearing by chance are stupendously astronomical; and the facts appear to support his calculation. The consequences of such statistical annihilation of chance (and therefore a purely materialistic explanation of life) are developed in Chapter 15.**[110]

Cosmo-logic

So to materialism's heresy, cosmo-logic.

In the beginning was *NOT* chaos. Modern physics shows that mankind dwells in a finely-tuned universe. To be precisely fit for life it must have started in a way most orderly, specific, specially defined. *Fine-tuned by chance? If low probability together with specific definition indicate design, no chance!* **A universe fine-tuned regarding many parts might be construed as one of specific, irreducible complexity; and, if lacking any part it failed to work, of minimal functionality. This pair's efficiency can, we shall see, indicate the hallmark of design.**

---

[110] *SAS* Chapters 19-25.

Perhaps you can see where this is leading. Does it suggest holism or materialism's upside down?

Firstly, information is an entity that exists independent of matter. The information-centre you are personally familiar with is mind. Science that's materialistic must exclude what's immaterial; metaphysic isn't scientific everyone agrees. Since when, however, could a philosophical exclusion outlaw mind's primary, informative reality?

Secondly, although materialism axiomatically excludes an immaterial element, if such element exists (as Natural Dialectic axiomatically supposes) it doesn't have to bear the properties or serve the laws of physic. For example, its 'holographic', unifying and yet 'nuclear' capacity for information may impact each 'pixel' (or each 'quantum') of the cosmos simultaneously.

Thirdly, as regards the order of creation[111] physics follows from psychology; physical succeeds the metaphysical; universal matter is projected through its mind. **In Dialectical terms, potential matter precedes physicality; it has been identified as archetypal memory.** Memory's the fixed, 'solid' or 'interminably repeating' phase of mind. The invisible reality of archetypal files constitutes an 'immanent potential'; **universal mind is the *implicit* universe.**

Mind's precedence is an informant whose principles (i.e. specific possibilities) are realised in the expression of particular physical phenomena. **The latter constitute an *explicit* body called our cosmos.**

Thus archetype is an object, thought or event's ***intrinsic being***. It is physics' 'hidden unvariable'. *Quantum physics is viewed as a phase in the realisation of archetype that yields its determinate physical 'overlay'. It represents a mid-step in the translation of universal mind into locally projected matter.* **As Natural Dialectic's framework clearly shows, quantum is the world <u>becoming</u>. Through its 'level' general potential everywhere is crystallised into what <u>have become</u> specific, local, physical realities.**[112] **Archetype, quanta and particular fixations are thus simply understood as, according to the three cosmic fundamentals, three different facets of the same reality.**

Accordingly archetype, that's metaphysical, emerges as creation's 'hidden invariant'. As definite notes in song or different rainbow lights so definite and changeless 'colour-coding' underlies the ever-changing stream of physical transactions. Perhaps cosmic memories are first physically expressed as various immaterial force-fields. Sufficient excitation of such field (or latent file) results, of course, in action of its sort; actual examples of abstract archetypes, characters called particles, appear. We're into physics' 'upper', quantum level now. **Thus order,**

---

[111] Chapter 12.
[112] *figs*. 3.3, 8.3 (with following stack) and Chap. 10: stack 2 may help explain.

both determinate and indeterminate in style, is just the same; according to this scheme of things these two qualities combine.

This conclusion is important to the point of repetition. Why should rules of physic govern metaphysic? A physicist's considerations are material but, if 'metaphysical linkage' does not have to cross physical space-time, it becomes instant. Archetypal influence is instantiated everywhere. This influence would be a property of 'pre-space', that is, of potential matter's void. Potential matter's been identified, unconscionably by naturalistic reckoning, as subconscious mind.[113] **Subconscious mind is a generic term for files of memories.** Could omnipresent connectivity reside in mind - not individual but universal mind? Not conscious mind but archetypal memory? **Thus logically quantum physics, dealing with the subtlest, fundamental forms of matter - force fields, pure energy and charge - would interface with metaphysic.**

Can Mathematics Help Us?[114]

Pythagoras, at least, believed that 'All is Numbers'; such immaterials help describe mind-matter frontiers whose formations are called archetypes. For others, including Roger Penrose, the Platonic world of mathematical forms is also real. He writes, 'There is a very remarkable depth, subtlety and mathematical fruitfulness in the concepts that lie latent within physical processes.'

No objects bear a number yet you number them and count. Numbers, symbols and mathematics aren't a physical but changeless, metaphysical reality. *Maths is a form of metaphysic.* Therefore, could essential, physically-independent numbers really govern physical complexity? Einstein, Planck and Eddington believed that, once you dropped upon their key, you could *deduce* (*top-down*) the reasons for all natural laws; you might unlock the codes whose inmost mysteries reveal just how a stable universe is sparked; a feat of mind *par excellence*!

Why, for example, are the electron-proton mass ratio, the fine structure, gravitational and other constants exactly what they are? Their formulaic outcomes are so physically co-operant you might proclaim that mathematics in the cosmos lets us live!

Symbolic numbers obviously pre-date humanity and they'll endure as long as cosmos but is this enough? Numbers are eternal abstracts yet it is energy not immaterial mathematics breathes the real fire; flames and motions not equations drive the cosmos forward. *Like the 'laws of nature' ciphers may describe but don't per se create a thing.* **Indeed, they have no being but in mind.** They inform the various quantities and qualities of mass and motion with a web of metaphysical relationships, transcendent

---

[113] see *figs.* 7.3 and 11.2.

[114] *SAS* Chapter 7.

logic and eternal principles. And if mathematics forged an archetypal link beyond the bounds of space and time it would, like any law of logic, have to be an abstract entity. In what could abstraction be embedded on a universal scale? For Natural Dialectic the reply is simple. **Mathematics is a real form of metaphysic; archetypes are also metaphysical and have no being but in mind. Maybe 'natural mathematics' is describing real, archetypal files.** Perhaps archetypes compose the link between the corners of what Penrose has referred to as a mysterious triangle of physics, maths and mind. At any rate, their logic's fixed. Fixed forms of mind are memories; thus they are memories in universal mind.

Towards a Theory of Physic and Metaphysic

You may understand, this way, that Metaphysical Reality yields physical appearance. Archetype orderly informs the expression of a physical polarity most basically expressed as energetic purities (forces) and constraints (particular things). Wave and particle, Bohr and Einstein, quantum and classical are each right - but, not unnaturally for physics, science misses out the Prior, Informative Reality.

Could *you* plan a cosmos better? Lucid physicists agree it looks 'as if' designed. **When it comes to astrophysics and cosmology stunning ingenuity seems to have coordinated chemistry and physics' natural laws, not least when it comes to bio-friendliness.**

Physic *and* metaphysic? *Metaphysic seems, except for logic, reason and mathematics, irrelevant to an operational description of material phenomena.* It is apart from physical events but not, perhaps, from their foundation. Unseen, it could be part of operational pattern; it may be relevant to *why* the system works, that is, why it issued in the form it has. It's worse than that. *Scientific materialism that excludes metaphysic essentially, delusionally excludes the instrument of its own operation - mind!*

The Dialectic's metaphysic rests with information; and for physics information is conserved. What kind of informative conservation is that but so-called 'natural law', the behaviour of energy projected from outside the edge of space - where 'outside' actually means emergence from an inner metaphysic and not some outer spatial or temporal place.

Where, therefore, is the projector; what's the transcendent nature of projection's source? Natural Dialectic's 'holographic' edge is everywhere; it's 'super-posed' on physics, omnipresent but invisible because it's metaphysical. <u>**It is the place where theme is turned to individual instance, principle is practised and where physic with its metaphysic meets.**</u> Space, time and things are *within* universal mind. **Mind's archetypes project our world; archetypal memories, potential matter, are the essence of our physic; they inform, unchangingly, material being. Immaterial information holds the world, physical and biological, together. It is by archetype conserved.**

Archetype as origin of natural law is not a new idea. Potential matter is sited on the interface at which subjective and objective meet. The metaphysical precursor to all physicality is known, in either individual or universal mind, as archetypal memory. Through this unconscious gateway are expressed the agents (quantum behaviours) and the outcomes (gross, sensible materials) of a starry circumstance that we call 'home'. Order therefore emanates from inside; guidance of the cosmos issues, hierarchically, from above. **The challenge is to integrate an archetypal form of information with physics as it stands today.**

The Principles of a Unified Theory of Matter

It might, in this cause, be said that Natural Dialectic simply takes what's there and reassembles it in an internally self-consistent way to make post-modern sense. Or again, that it tracks along its sub-routines to improve sense of the cosmic program as a whole. *Every systems analyst understands top-down logic.* If you want to understand how, *top-down*, anything works you have to begin at the beginning; you have to understand its purpose and its principles. **For a full understanding**

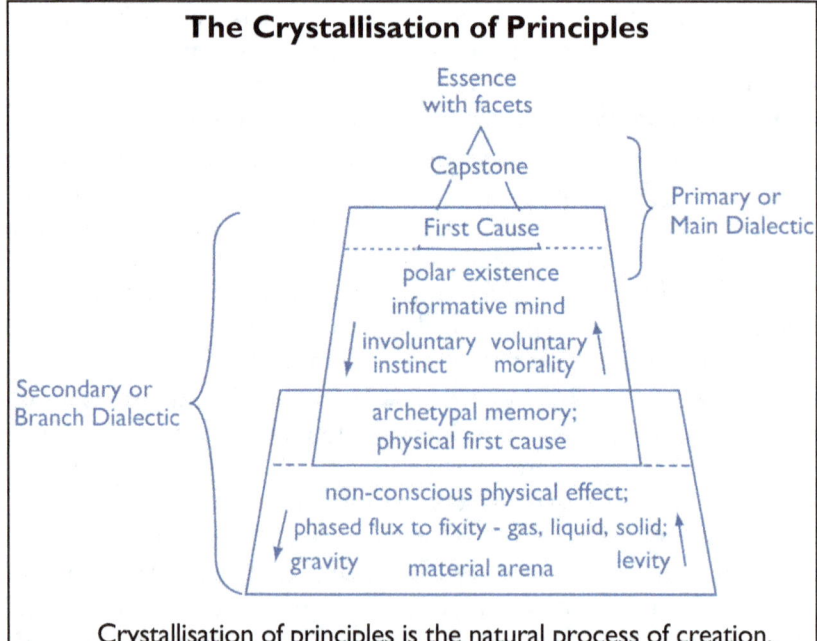

Crystallisation of principles is the natural process of creation.

Physics inspects the world's non-conscious stage. It examines what's 'below' the physical first cause. This cause, dialectically 'wedged' between mind and matter, is an archetypal step reflecting, as potential matter, the Capstone's higher, existential First Cause.

fig. 14.1 (see also figs. 7.2 and 7.3)

you need to grasp first principles. To grasp the principles behind physics you may need to consider their metaphysical origin.

What is a principle? **A principle, which has been called a highly condensed form of information, is conceptual.** Because it precedes, orders and directs subsequent and often complex actions, *principle is potential action*. Because it is an apparently absent latency, nothing without practice, principle is *symbolic action*. What is law without behaviours, idea without fruition or intention *sans* its tool to implement? Principle orders action. It is the source and definition of all legal activity. Its intent is expressed through mechanisms that accord with its logic. *This is no less true of the cosmos than the latest plan or program of action (however grand or trivial in scope) that you brought to fruition.*

Principle informs practice. For this reason it has been argued that active precedes passive information, concept precedes manufacture, thought precedes action and, axiomatically, mind is the repository of principle. In other words, mind is the potential for material order.

Because they reflect the way principle is (or could be) translated into practice abstract mathematics and the laws of logic can predict and compute automatic physical relationships and behaviours. Physical principle, the 'instinct' of nature, gave, gives and will give rise to matter in a specific order and in order to understand creation properly it is imperative to follow this sequence. The narrative must follow its logic. *The right order gives the right reasons and, from these, the right results.*

Top-down, principle precedes practice and the influence of principle is increasingly 'crystallised' through sub-principles, ideas, archetypes and, finally, physical phenomena. Therefore, one needs to understand the origin of principle to understand its consequent practice.

Before attempting to rephrase scientific jargon, consolidated perspectives and speculations in terms of Natural Dialectic, it is necessary to paint a broad-brush sketch of what sort of principles and conceptual norms that a holistic mind-set (involving elements of energy *and* information) might include:

1. *top-down* scaling; a conscio-material gradient
2. interplay between cosmic fundamentals in different proportions in different cases
3. Archetypal First Cause whose essence creates mind; and, at the base of mind, a physical first cause that constitutes the archetypal norms to which material patterns of behaviour cleave. In any actual context - psychological or physical - first, nuclear causes 'germinate' the rules by which events occur. Such informative potential is identified with the tier above its expression. In dialectical terms the informative tier above matter is mind; and in mind potential matter constitutes the lowest and sub-conscious tier. This implies that the principles governing the behaviour of material energy (the scope of physics and

chemistry) are lodged in and derive from a metaphysical source. **We make sensible, sensory connection with matter; but we make contemplative connection with its metaphysical principles**.

Thus, in principle, the order of a simple argument might run:

tam/ raj	*Sat*
lesser truths/ appearances	*Truth*
aspect of reality	*Reality*
sub-principles	*First Principle*
particular expression	*Principle/ Law*
local practice/ outcome	*Potential/ All Possibilities*
subsequent order	*Archetype*
elaboration	*Plan*
change/ inconstancy	*Stability/ Constancy*
periphery	*Centrality*
polarity	*Neutrality*
↓ *tam*	*raj* ↑
descent ←	→ *ascent*
from Axial Source/ Hub	*towards Axis/ Hub*
separation/ > duality	*unification/ > unity*

for mind (information-dominant; ↑ **truth-seeking) we write:**

tendency >passive/ automatic/ lifeless state	*inclination > purposive/ meaningful behaviour*
towards infra-consciousness/ sub-consciousness/ oblivion	*> ultra-consciousness/ super-conscious awareness*
> non-conscious physicality	*> Natural Heart*
inc. appearance of coincidence	*inc. perception of order*

for matter (energy-dominant; ↓ **non-conscious) we write:**

increasing loss of free energy/ entropy/ bondage/ fixity	*energetic gain/ negentropy/ flux of process*
resistance/ inertia/ exhaustion	*action/ motion/ freedom*
repetitive motion/ immobility	*stimulus/ change*
grosser/ materialisation	*subtler/ dematerialisation*
gravity	*levity*

First cause physical or potential matter correlates with fig. 12.2's upper syntactical level of passive information. It is also called transcendent potential, absolute matter or archetypal memory. Having identified the First Principle (and also Principal) of a Unified Theory of Matter we have discussed the immaterial nature of its material absence, that is, the immanence of its 'pre-physical' void.

The order of play now turns to its (*raj/ tam*) vectored physical issue - the lower syntactical level or lowest, quantitative level of passive information. This level is the final expression of principle in physical practice, that is, in terms of data items called objects and interactions called events. It involves[115] two sub-divisions.

(Raj) Active Energy

Active energy is (see Chapter 11) known to Natural Dialectic as **matter-in-principle**. Physics deals in detail with its quantum and atomic physics[116] whose simple, primary agents give rise, in congregation, to subsequent complexity.

(Tam) Passive Energy

Passive energy is known as the classical phase of bulk matter or **matter-in-practice**. Such matter is 'bonded' and includes molecular formations, plasmas, gases, liquids, solids and, of course, all study related thereto. It represents projection's furthest radius from Source.[117] The subjects of physics and chemistry deal exhaustively with this level of creation.

How, though, does science deal with its own non-physicalities? Surely space and time are these?

Space

Archetypal memories are nothing physical. Concepts 'frozen' in a universal mind, they do not exist in physic's space-time. But the logic of Natural Dialectic sees (*tam*) physical patterns of behaviour as the low-level, non-conscious, automated consequence of mind - instinctive mind that, as its internal root, governs involuntary, 'crystalline' matter. This governor's body is our astronomical universe. Physics and chemistry study the actions of its 'instincts' or archetypal control systems. These are natural law and their 'nervous system' is actualised in electromagnetic and other fields of force that orchestrate the way things interact. **In this view, contrary and outrageous for materialism, matter is developed memory; it is a projection.** And, at the outer rim of this projection, energy is frozen; any solid represents creation's edge; it represents, full stop, world's end.

The immaterial is indeed, by definition, irrelevant to material science and anyone who demands physical as opposed to inferential

---

[115] *fig.* 7.3 section 5.

[116] Karl Popper dubbed quantum physics 'the transcendence of materialism' but for Natural Dialectic this phrase applies to the *metaphysical* element called potential matter/ archetype.

[117] *figs.* 8.2 and 14.1.

proof of non-physicality may well end up denying even his own mind. So let's retreat. Let's make space for space and turn to scientific nothingness in terms of space and time. What is the nature of these immaterialities? Are they metaphysical or not?

	tam/ raj	*sat*
	business	*potential/ latency*
	action	*source*
	archetype-in-motion	*archetype*
	polarity	*neutrality*
↓	*negative*	*positive* ↑
	gravitation	*levitation*
	contractive	*radiant/ expansive*
	isolating	*relating*
	mass/' dead appearance'	*'live stimulant'*
	sink	*causal agency*

Top to bottom exhaust it. Freeze your black box, pump out every particle and piece of radiation to get nothing in return. A bare vacuum. Material immateriality. Objective but inert, impotent 'metaphysic'. Is outer space not such a box without its sides?

It isn't now! Nothing isn't nothing when it's virtually something! And that's not all. Modern space bends, contracts, is grainy and, expanding with a levity (dark energy or anti-gravity), carries cosmos forward![118] It just isn't perfect. It is 'false' and full of fields; it teems with gravitational, electromagnetic and (uncertainty is certain of it) virtual fluctuations; and *ZPE* (zero point energy) is flickering with in-and-out-of-being particles.

In this case it has been proposed that vacuum might comprise informant source as well as energetic sink. Indeed, unless some instability broke cover, might not balanced tensions stay unseen? In theory, 'virtual' quivers stay subliminal but sufficiently excited fields give rise, spontaneously, to quantum particles/ actual energies. And just as Pauli's distinct quantum states give rise to atomic structure so, it is suggested, distinct, specific fluctuations of a vacuum 'harmonically' give rise to fundamental particles and forces.[119] This is the graduated density of vacuum and the nature of a cosmic egg; preordination rules; polar, fundamental (↓↑) vectors specify material types; a scale of immaterial fields constrains the action. Measurable patterns would appear. Then space is seen as a 'container' and the active vacuum as a large-scale latency with 'immaterial fields for development'. The world is thence developed on such filmy '3-d template'. *ZPE* and 'virtual photons' mediate all electromagnetic interactions. They are postulated as support for an electron's form, atomic structure and thence the

---

[118] *SAS* Chapters 9 and 12: Pass the Paracetamol.

[119] Chapter 13: Psychosomasis.

phenomenal universe. The appearance and disappearance of clouds of virtual particles in this very lively vacuum are a price paid for what is claimed to be the most mathematically precise theory ever known - Richard Feynman's relativistic quantum field theory of electrodynamics.

Does archetypal symmetry with neutral space devolve the world's polarities? What is Natural Dialectic's spin?

Its vacuum's paradoxical. Vacuum is essential both as start *and* end-point, source *and* sink. It is (can you see?) more than single. It's a trinity. **Firstly, (*sat*) void is an essential form of nothingness; non-physical, it is an archetypal 'emptiness', a pre-creative latency.** For mind the Archetype is *Logos*; but for matter it is archetypal memories, potential matter or the principles of natural law established in the metaphysical substrate of universal mind. Law is the frame of possibility; its activation yields specific, orderly events.

Next, (*raj*) microscopic, kinetic space comprises fields in activated but subliminal or fundamental mode, that is to say, 'virtual' and actual fields of radiant (levitational) and contractive (gravitational) events. *As well as negative sink it is suggested that, in positive aspect, quantum void is a source of perpetual, supportive energy channelled into quantum forms.*[120] **Such a reservoir of potential energy would thus constitute a medium between matter and its immateriality.**

Finally, (*tam*) macroscopic space is what appears inert, oblivious and empty. It seems, in this aspect of absence, that there's nothing there. There are apparently no things at all - except, perhaps, dimension and extension. Dead void! Full impotence! Material immateriality! Such, physically, is the passive emptiness of what we know as 'outer space' - though atoms of the world and their bulk aggregations are, since space takes more than 99.9% of their own volumes, also hugely full of void; without your space you'd be invisible - very heavy micro-speck of dust!

In summary, vacuum catastrophe.[121] (*Raj*) quantum and (*tam*) relativistic appreciations of emptiness hugely conflict. The latter's space is 'thin'; its explicit phase is bare but, as physics notes, the former brims full with implicit energy (such as *ZPE*). **Its 'kinetic' species of space (that is notionally nothing) may well substantiate, in its various aspects, the form and action of material phenomena.**

Time

Yes, you can think of space and time together. *Space-time (if you really have to tie the knot) is the means of separation, differentiation, isolation without which nothing could appear.* Without the duo's continuity how

---

[120] *SAS* Chapter 9: Nothing.
[121] *SAS* Chapter 8: Holy Grails.

could apartness happen? Without such isolating yet concatenating continuity there would be, literally, neither space nor time for anything; without these differentiators nothing moves or stands alone; without these dimensions what a sudden cramp would clamp, what a paralytic lock-out of the most impotent kind would leave you neither jot nor dot of anything. Nothing. Nothing, even these, a couple that are nothing-in-themselves!

Or you can think of time apart. Is time's substance motion? So that its reality is change. If things change does changeless, inert space have time? Time is spaceless; is space timeless? Is time everywhere and nowhere all at once? And is time as flexible according to velocity as space is gravitationally bendy? What is it actually bends or flexes if there's nothing there?

Time is a powerlessly powerful absence from materiality. You can phrase it several ways.[122]

Think, for example, of materially graded time. A motif that underlies the presentation of Natural Dialectic is one that reflects the gradient of creation. *The gradient runs top-down.* Cosmic hierarchy drops from mind to matter, from archetype to physical expression. Principle (condensed information) guides practice; thus simple runs to complex, general to specific, universal plan to detailed, individual, localised expression.

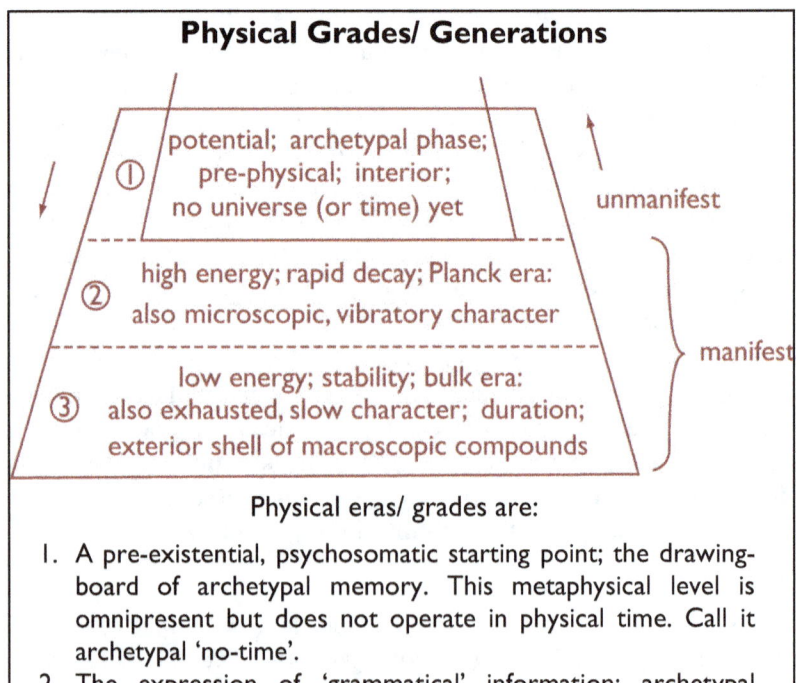

Physical eras/ grades are:
1. A pre-existential, psychosomatic starting point; the drawing-board of archetypal memory. This metaphysical level is omnipresent but does not operate in physical time. Call it archetypal 'no-time'.
2. The expression of 'grammatical' information; archetypal memory is expressed as a sub-atomic, quantum or principle

---

[122] *SAS* Chapter 10: Time.

> grade of matter (represented in the Standard Model by three generations of fundamental particle); this level amounts to nature's intrinsic code for matter; its dynamic alphabet operates in 'fast' micro-time and may be construed as the level of quantum activity.
> 3. Nature seeks the lowest energy-level. This turns out to be, by now, the last-occurring generation of matter (ours). Its particles (electron, neutrino, proton and, inside an atomic nucleus, the neutron) endure. Their atomic border gives way to a multiplicity of differences in the form of plasma, molecular gas, liquid and solid. The 'internal' energies locked in these low, classical grades of matter are vibratory; and external definition increases with loss of energy-in-motion. This level of time, in which you and I live, operates in 'slow' or 'sensible' macro-time. Call it 'long or slow time'.

fig. 14.2

	grade	time
sat	pre-physical latency	-
raj	quantum micro-level	Planck/ quantum era of high energy and quick time
tam	classical macro-level	era of low energy/ bulk aggregates and slow time

If grades run from potential to exhausted then hierarchy is both metaphorical and real. Physics' cosmos is creation's rim; central metaphysic is within. *In this sense (figs. 7.1 and 11.3) it follows that higher is 'within' lower and first is 'within' last. What comes first informs what follows. 'Higher' is the magisterial cause of 'lower', consequent effect.* Each grade nests within the one below. Atoms are nested within bulk materials, particles within atoms and vacuum within everything.

The clear implication is that creation springs 'organically' from within outwards. The dialectical expectation is, therefore, that subdivisions of the physical cosmos devolved in the same order in time as in material grade. Creation moves from centre towards periphery; 'top' or potential matter transcends its expression as, firstly, quantum and then large-scale, classical matter.

Alpha Points

Every object and event, every pattern and behaviour has a birth, lifetime and death. For you, earth-time began at your conception as, for certain, it will end upon your body's death. However, has our human intellect capacity to fully grasp the birth of cosmos or demise of its non-conscious energies?

Bottom-up, are cosmic changes rung eternally? Or did they issue from a most mysterious big-bang?[123] Will speculation differ in five hundred years?

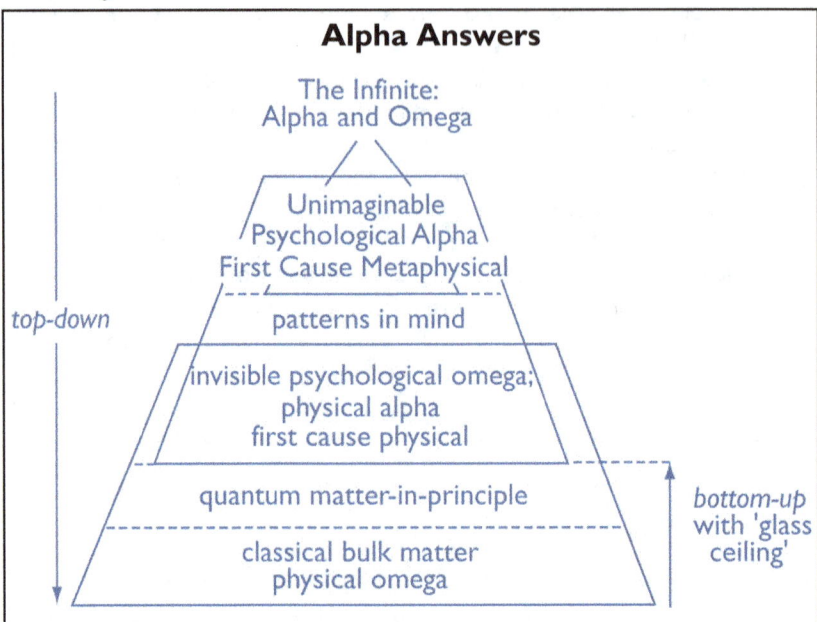

Alpha Answers

Top-down the Source is Infinite. In this view a 'glass ceiling' (made, maybe, of vacuum energy but certainly inclusive of transparent space) is reached where metaphysic's transformed into physic. The pattern of physical energy is issued 'through' this 'glass ceiling'; in other words physical derives its origin and sustenance from metaphysical.

Bottom-up the *top-down* view is absurd. A philosophical decision has been made - there's nothing that, even prior to cosmos, is not physical. Logically, therefore, everything physical must have had something physical precede it - except perhaps at first. There's no glass ceiling, nothing is unreachable by naturalistic means. Things must, of course, have sprung from nothing physical but 'nothing-physical' is not, in this view, metaphysical. Therefore the natural, material universe must be derived from wholly powerless emptiness.

Which is the A-grade answer? Which the E?

fig. 14.3

Top-down, most likely, as we saw,[124] our starry universe was metaphysically kicked off.

---

[123] *SAS* Chapter 12: Magnificent Mythology.
[124] start of this Chapter.

Energetic cosmos is, we say, (↓) entropically or gravitationally inclined.[125] Vast initial input of free energy is 'running down' but stays, in quantity and qualities, the same. Through all creation's changes particle and force remain non-consciously the same - involuntary, oblivious. Such oblivion is automatically locked in its own loop.[126] No will, no inspiration or awakening mean no escape but which proton or electron ever yearned for liberation?

The tendency of mind is, on the other hand, (↑) negentropic.[127] Its informative drive is (using principles like symmetry, comparison and connectivity) to question, understand and problem-solve. Its ascending loop may consciously reach towards Mount Universe's Peak; it may enthusiastically expand towards Knowledge of Full Truth. In realising Source it understands all nature's Alpha Point.

Points Omega

	existence	*Essence*
	created events	*No Thing*
	end/ action	*Initiation*
	sink/ process	*Archetypal Source*
	polar expression	*Potential*
	alphabet	*Alpha = Omega*
↓	*informed*	*informant* ↑
	entropy	*stimulation*
	end/ death	*process/ lifetime*
	exhausted void	*action*
	omega	*alphabet*

Initiation, action, end: this is the omnipresent dialectic order that pervades creation. Things were originally wound up but how at last, on winding down, will they wind up? It's time to end the universe of physics.

The business of annihilation hangs upon your standpoint. Was there steady-state, eternal matter, an uninitiated world without an end? Or was our cosmic missile issued from a non-existent gun? Was it mythic, multiversal phantoms or titanic branes whose clashing sparked off worlds?

And, if there's an end, will expansion cycle with contraction in a yoyo-go - perhaps even differentially as various local masses crunch by (↓) **gravity** at different rates (and therefore times) into their so-called 'infinite black densities'; or will all vanish down One Great Black Hole? On the other hand, could (↑) **levity** define the nature of point omega? Will creation's empty womb accelerate its spawn to death; must the lights go out inside a side-less, barren and expanding, everlasting tomb?

---

[125] Chapter 4 (*tam*) cosmic fundamental; Glossary.
[126] *SAS* Chapter 18: Death.
[127] Chapter 4 (*raj*) cosmic fundamental; Glossary.

As usual, holistic hierarchy changes everything. If Alpha Silence speaks what happens when the speech is done? If latent archetype of mind (*Logos*) or physicality (potential matter)[128] is vibrated into action what would happen if the excitation ceased? Formlessness projected form and, when power is cut, forms collapse through stages whence the grid had been initially devolved. Instant chop. Universal power cut. Back to Silence. Dissolution of existence means that only Essence stays. Once the roll of *Om* has died away, Alpha Source and Omega are revealed one and the same. This, the Source of nature, is Most Natural.

The Matrix

A scientific culture's one of healthy doubt. Is faith in doubt a working premise or a final attitude towards life? Could such a culture bring itself to doubt its own foundations? *In this case would a dose of doubt about materialism do?*

The dose might come from east and west. We could check what Einstein, Bohr, Heisenberg, Sir Jagdish Chandra Bose and Satyendra Nath Bose FRS believed. Sir James Jeans noted 'the universe seems nearer to a great thought than a great machine'. But let the final word rest with Max Planck. Planck not only first read, recognised and published (in the *German Annals of Physics*) the start of Einstein's revolution, the latter's Special Theory of Relativity. He also pioneered quantum theory, the second pillar of modern physics whose study is sub-microscopic, sub-atomic phenomena - the matter-in-principle of Natural Dialectic. Actually, his foresight may have ushered in the next revolution in human understanding which has already begun to focus on the primacy of the 'unscientific' informative co-principal[129] as opposed to its energetic coordinate. He asserted that the discovery of truth can only be secured by a determined step into the realm of metaphysics and, at a lecture in Florence (1944) called '*Das Wesen der Materie*' (The Nature of Matter),[130] said:

"As a man who has devoted his whole life to the most clear-headed science, to the study of matter, I can tell you as the result of my research about the atoms, this much: *there is no matter as such.* All matter originates and exists only by virtue of a force which brings the particles of an atom to vibration and holds this most minute solar system of the atom together... We must assume behind this force the existence of a conscious and intelligent Mind. This Mind is the matrix of all matter."

Indeed (The Observer 25-1-31 p.17), "I regard consciousness as fundamental. I regard matter as a derivative of consciousness."

Perhaps Max Planck would have appreciated Natural Dialectic.

---

[128] *figs.* 7.2 and 14.3.
[129] Chapter 11: Informed Energy; also *SAS* Chapter 2 First Principles.
[130] https://en.wikiquote.org/wiki/Max_Planck

Chapter 15: Triplex Biology

A life-form is composed of energy that is highly informed by programs derived from complex, specific code. Biological forms therefore show evidence, like technological and computational machines, of purpose and intelligent design.

Who denies your body is a fragment of explicit, cosmic body? And your intrinsic human being is implicit; it is an archetype in universal mind.[131] The three levels of biology thus follow the same as those of mind and matter. They descend, from internal externalising, in the order of the cosmic fundamentals.[132] Therefore they run from (*sat*) potential matter (archetype) through (*raj*) energetic, metabolic phase to the (*tam*) 'locked', classical phase of bulk structures.

(Sat) Potential Biology

Potential precedes possible action.[133] It is a prerequisite or precondition for results but, as noted in previous chapters, Natural Dialectic does not think of physical potential in the way that physics does. Potential matter is a metaphysical affair.[134] In biology, informative potential is laid into programs for irreducibly complex, minimally functional mechanisms operative within cells and organisms.[135]

(Raj) Active Biology

Active biology occurs at the level known to Natural Dialectic as **matter-in-principle** - quantum, atomic and molecular. Here find the nanotechnological communications of *H. electromagneticus* (Chapter 13) and the codified operations of electrons, protons, atoms and molecular biology. The primary agents of code and metabolism give rise, in congregation, to subsequent bulk functional and anatomical complexity.

(Tam) Passive Biology

Passive biology is the subsequent level of subcellular machinery, cell

---

[131] *figs.* 13.5-7, Chapter 13: Frozen Time to end of Chapter; Chapter 14: Cosmo-logic; also *SAS* Chapters 15-17, 19 and *AMA?* Chapter 24.

[132] Chapters 3, 4, 7, 11; also *SAS* Chapter 1.

[133] in Dialectical order 'potential matter' or 'archetypal phase' was referred to both in Chapter 11 (under 'passive information') and in Chapter 12 (under 'information's infrastructure - code') as comprising an *upper linguistic/ codified or syntactical level*; and items of the two lower, physical phases (active quantum and passive bulk matter) were similarly collected into the *lower linguistic/ codified or syntactical level*.

[134] see *figs.* 7.3 and 8.2; also Chapters 9-11; see also Glossary: archetype.

[135] *SAS* Chapters 20, 21 and 23; *A&E* Chapters 9-11.

type, tissue, organ, system and the phenotype of organism. Until the explosion of biochemical and molecular studies this, its classical phase of bulk matter or **matter-in-practice**, was the sole focus of biologists.

For the purposes of this chapter, however, we'll focus particularly on the first, potential stage. In our current age of information technology we shall concentrate on the origins and operation of bio-code.

The Basis of Biology is Information

Biologist Theodosius Dobhzhansky is famous for coining a popular mantra: 'nothing makes sense in biology except in the light of evolution'. *This is utter nonsense.* **The actual, iconoclastic fact is: 'nothing makes sense in biology except in the light of information.'** You may drag evolution in on information's tail but it is simply a fashionable word (check carefully in the Glossary) used in biology to mean several different things. *In reality, codes and signals run the bio-show.* **Not evolution but information is the basis of biology. Thus the subject's basis is an immaterial entity. The subject's root is metaphysical.** *To recapitulate: it is not, trivially, an origin of species but, fundamentally, the origin of information that is at issue.*

communication	*Concept*
expression	*Archetype*
appropriate contextual action	*Informative Potential*
↓ *informed product*	*informant code* ↑
lower linguistic level	*upper linguistic level*
physical item/ body	*bio-logical grammar*

Information's sprung in mind and mind recruits some kind of symbolism to materially express itself. Symbolic code[136] includes formatting rules (such as language with syntax and grammar), carriage (using sound, light, electrical activity and so on) and storage (memory, tape, books, *DNA* etc.). It includes communication whether through inanimate or animate an agent. Organisms involve informative agents in abundance. Mostly their intelligence is reflected passively, as machines reflect intention; but active, creative, conscious response occurs in some as well.

From this point of view matter has no chance of ever generating mind's core *sine qua non*, information. As Chapters 11 and 12 describe in detail, active precedes passive state. Mind's creativity rules matter's lack of it; mind's message gives rise to material patterns which, according to its purpose, other minds can understand. This frame of reference treats biological organisms as passive or 'rigidified' expressions of information.

---

[136] Chapters 10 and 11: code (**upper linguistic level**) and informed product (physical item at **lower linguistic level**); also *SAS* Chapters 5 and 6.

Again, therefore, the basis of biology is metaphysical.[137] Metaphysical leads, physical follows. The correct angle from which to view biology, whereby its parts fall into place, is therefore from the angle of immaterial information. Code, syntax and semantics - translation to and from a purpose - are the operators of information. The origin of life is abstract code. Whence appears, in all the universe, by agency of wind or heat, dust or rain, a set of abstract symbols? **A cell is seen as information first, chemicals second.**

The Principles of a Unified Theory of Biology

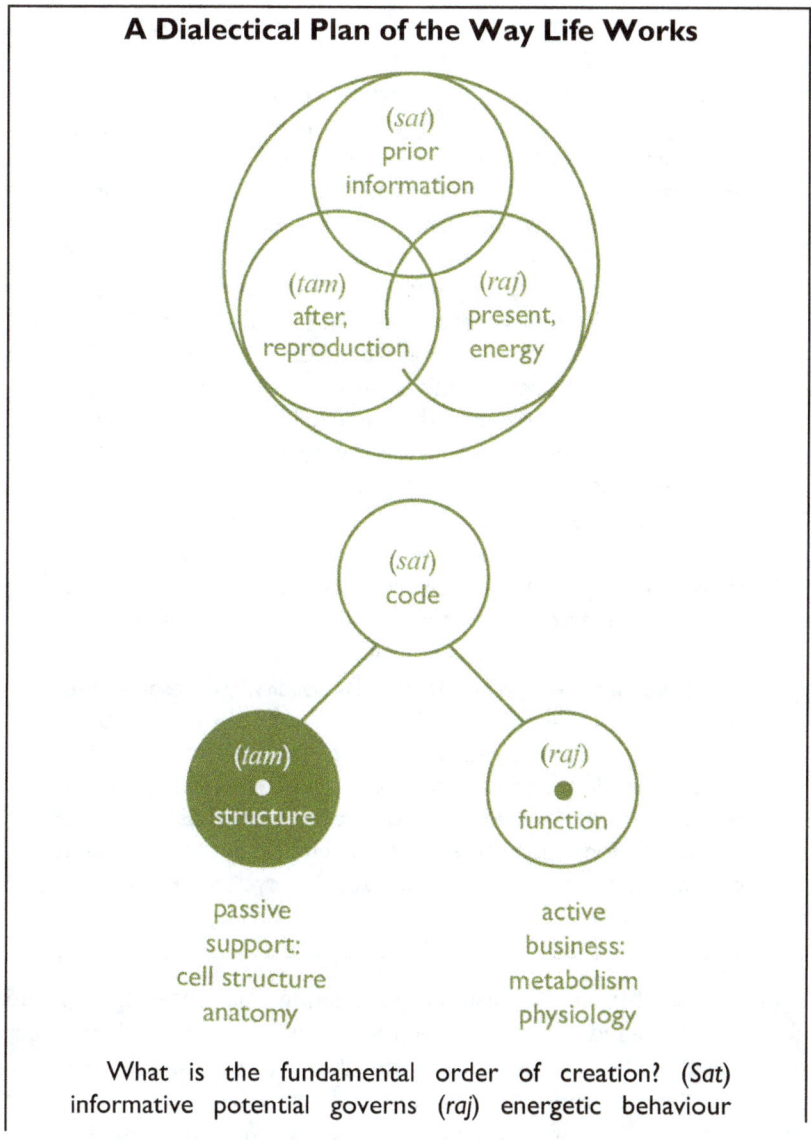

[137] *SAS* Chapter 19.

> whose action lapses to (*tam*) exhaustion. Biology, we'll see, is permeated by such order.
>
> *Information precedes.* It is prior and anticipates. Information is the potential and *sine qua non* for behaviour that consequently issues orderly.
>
> **Information comes, as previously explained in Chapter 11, in two forms - active/ conscious and passive/ unconscious.**
>
> The **conscious, informant case** involves sensation, intention and knowledge; it employs, in order to experience the body-world, a voluntary nervous system.
>
> The **unconscious, informed case** involves both psychological and biological components. These include archetype (involuntary instinct and morphogene); reflex balance, called homeostasis, by way of nervous, hormonal and other systems; cybernetic metabolism controlled by preordination in the form of genetic code carried chemically by *DNA*; and muscular organs of action and response. Life's process is one of dynamic equilibrium. Equilibration. Its goal is balance in accord with pre-set norms. Metabolism, being totally information-dependent, works with reference to precise, incoming messages and equally precise genetic response. Such response is indexed, switched and flexibly monitored by non-protein-coding and epigenetic factors. Life is, in this way, an incarnate flux of order due to information.
>
> **Energy** provides for survival now. It involves, metabolically, photosynthesis and respiration. It promotes cell biochemistry, trans-membrane voltages, physiological processes and, on the large-scale, (nervous) sensation and (muscular) motion. The character of all function is energetic.
>
> **Structure**, whose character is solidity, represents the outermost, fixed (or flexibly fixed) realisation of shape. Its base domain is energy's container, a fixed expression of internal, orderly flux. In other words, a 'phenotype' (see Glossary) is a peripheral aspect whose body both reflects and fixes the shape of inward information and energy. The end-product of structural development, maturity, is reproductive. The cycle starts again.
>
> *fig. 15.1*

Fundamental principles instruct the functions and encasing structures of biology; and complex information systems, hierarchical communication, homeostasis, energy metabolism and systematic structure (including the mechanisms of exhaust and reproduction) are basics that compose the outline of all organisms. *How, when these parameters are needed from the start of any life, could nature creep towards complex and successful coordination out of accidents?* **It may be clearly demonstrated that the panoply of life's fully informed co-operations is as entirely consistent with deliberate design as it is inconsistent with neo-Darwinian evolution's core 'co-creators'.** *This couple comprise non-orderly (that is, chaotic) chance and non-creative selectivity by death. Such mindless co-creators can't, we'll find, create a single cell.*

Natural Dialectic is one way to phrase biology. You could equally apply the reasoning of an engineer, that is, informed, intelligent construction. Systems generated randomly don't merit rational analysis. No-one denies that information runs the bio-show; nor that, as with machine, form and performance of bio-machinery appears 'as if' deliberate. We know that all technology and printed literature passively incorporate intelligence.[138] Mind (of their creator) lives invisibly and inextricably in them. But, by itself, can sheer matter (even grey stuff) comprehend, innovate and system-build? Or codify its aimless plans with script that turns into a purposeful reality? We're here, it is agreed, but should that mean we *must have* happened accidentally?

A book's as unintelligent as ink and paper; so are specific gigs of molecules. Thus, *bottom-up*, material information *is* its mindless agents, that is, its chemistry of carriage. It is passive; it's been gradually, obliviously 'organised' by chance mutations acted on by variable constraints. Natural histories are promoted to support this evolutionary view; plausible narration presses to explain such mindless source of bio-information; science promenades with Lady Luck. *A post hoc story that articulates a chain of possible historical events is, however, not a scientific exercise in the same sense as operational physics, chemistry or applied biology.* Is validity exclusively conferred just by taking a materialistic stance?

Conversely, *top-down*, mind cuts straight to chases mindlessness can never see; and, always, coded plans instruct resultant bio-form. Cells, with their nucleic acid, clearly operate like any automated factory; they embody information 'by analogy'; it is 'as if' their core displays intent. Life, psychologically and biologically, is unequivocally information-super-rich. *When does appearance of intent become so overwhelming you infer it's real; whence does what's inferred to be intentional arise; what's the nature and the source of information and, to wit, bio-information?*

Is randomness informant or informed? Rather, it degrades and swamps constructive logic. Faith that the basis of biology derives from fecklessness of chance is blind. **Thus we have to argue that mainstream science (and, in particular, molecular biology) has yet to fully grasp the clear, informative implications of its discoveries.**

More and more molecular biology shouts metaphysic since, as well

---

[138] *SAS* Chapter 6: Machines, Mind Machines, Universal Authorship.

as the chemical hardware, it increasingly illuminates the signals, switches, codes and carriers of complex operating software.

The Central Executive is Homeostasis [139]

As well as code (*sat*) balance is the axis round which life is swung. The chief arbiter of healthy operation is, in every cell of every organism, a cybernetic set of checks and balances.

↓ *inertial equilibrium*	*Equilibrium*	*dynamic equilibrium* ↑
flat-line	*Target Mean*	*regular cycle*
feedback failure	*Pre-set Norm*	*negative feedback*
imbalance	*Axis/ Pivot*	*equilibration*
ill-health/ death	*Vis Medicatrix*	*vibrant health*

Life, the integrator, draws together; death, disintegrator, splits apart. Biological form waives, temporarily, the tax of death. The cybernetic instrument of this avoidance, which allows a body time on earth, is called homeostasis. Executive homeostasis is informed, hierarchical and cyclical. Whether psychological or biological, it means adherence to a norm. Such norm is its target. This may be psychologically flexible (as with changing desires) or biologically fixed (as human thermoregulatory control pre-set at 37°C). Either way, in maintenance or development, procedure cycles cybernetically around a stabilising norm. Cycle, wave, vibration - call it what you will. Such dynamic equilibrium[140] keeps (*sat*) balance and stability that we call health. Its vital control involves a switching system called *negative feedback.* Your home's central heating system is a simple example of such feedback. *It needs, along with associated plumbing and water supply, a minimum of three integrated mechanisms to work.* Firstly, a *sensor* tells the temperature. Secondly, a *regulator* decides if it is equal to, above or below a pre-set 'norm' and, finally, an *effector* accords (switches the boiler off, on or does nothing). Such a system, allowing the temperature to fluctuate, is, in effect, oscillatory.

It is worth emphasising that cybernetic homeostasis involving input, output and a balancing or regulatory processor has no use whatsoever on its own. It does not work in a vacuum but involves an irreducible minimum of complex, associated components and, in metabolic terms, pathways. Sub-systems have purposes and are, in cooperation, interlinked in a hierarchical fashion within an overall scheme or, in this case, biological organism. All codified factors, whether at molecular, cellular or bodily level, must therefore necessarily be fixed together in the right relationships at the same time in order to work.

---

[139] *SAS* Chapter 19: Biology is Hierarchical and Cyclical.
[140] see Glossary: equilibrium.

Tight-rope balance is the key. Each individual, homeostatic cog cooperates in balance with its overall scheme. This scheme may involve an individual body but also, in a wider sense, an ecological community. Where health is balance disharmony does not feel right, discord calls the doctor and cacophony cuts out health's song. Death's the trump of Old Man Entropy - except for a purposeful and advantageous form of death called apoptosis. Such organised cell-death promotes, like scaffolding, construction. How, you ask, could death-and-destruction harnessed as a developmental tool survive a piecemeal evolution of its highly integrated steps? More broadly how, as check-mate to extinction, could any codified development have cycled accidentally unto maturity?[141] Vital balance dances on death's tomb. It holds the end at bay so why, if life's end was unforeseen, did there evolve genetic patterns of senescence (such as the Hayflick limit) and, for every type of organism, preordained life-cycles? And, if homeostasis is the key, why hasn't a bio-generalised extension of survival-time progressed towards the evolutionary apotheosis of a mortal's immortality? Finally, why and how at random should blind molecules create an interlocking set of cybernetic programs that are vital for each living moment? Or was life's tango pre-composed?

In short, so central and fundamental is the informed balance of homeostasis to all life on earth that a visible life-form, whether uni- or multi-cellular, should be understood as, first and foremost, an invisible but preordained pattern of informative control. Such multiplex control could not evolve gradually or by accident - except by wishful thinking.

Nuclear Super-Computing

A computer is an information controller. It is a mind-machine entirely dependent on programs and, in the form of chips and databanks, memory. Its construction and encoded programs are teleology objectified. Computer operations are based on conceptual Turing machines. These consist, like homeostatic systems, of input, processing controlled according to a set of rules and output. Scientists (such as Leonard Adleman) have realised that *RNA* and proteins responsible for *DNA* manipulation act, with the *DNA* itself, as conceptual machines - input *DNA*, process, output specifically required expressions of stored code. Their biochemistry operates as a Turing machine; bio-information is computerised. From this realisation the scientific development of artificial *DNA* computing is being developed.

No ifs or buts. Biological computing and its issue, biological machinery, are not 'as if'. They're not illusions or appearances but real. Here natural's as conceptual as anything that's artificially engineered. In

---

[141] *SAS* Chapters 24 and 25.

this case the excuse of analogy, beloved of evolutionists, is swept away. Call them alphabetical or digital (as machine code) but *DNA* molecules store digital information; and the nucleic acid/ protein system *is* a calculating machine. Information, abstract and metaphysical, is carried on biochemical 'speech'. This 'speech' is subject to the linguistic architecture of alphabet, syntax and grammar; its expression is tightly regulated. Sentences, phrases and whole routines - metabolic modules - are fashioned to most efficiently exact desired effects, that is, requisite end-products for their system to survive. No doubt, as chemicals, genes are hardware but just like hardware called a c-drive or a book they carry immaterial meaning, reason, purpose and program. Informative message is carried on the concrete arrangement of materials. **Genetic code is as designed as sentences you think and speak.**

Bottom-up, the problem's to convince yourself that, granted a vastly complex starting-point, the computation of cell chemistry haphazardly, mindlessly 'improved itself'. Molecules 'progressed'. But what systems flow-chart could you mindfully construct to demonstrate this possibility?

Top-down, programmers know that, from a main routine, switches branch to sub-routines and, when a sub-routine is done the process cycles back to start again. **These routines are modules.** This conceptual character of algorithm, this repetitious use of switches and blocks of modular code is just what coded bio-systems show. It is how nature's life-forms, full of reason, always work.

DNA and *RNA* are nucleic acids; and, actually, these acids compose a **genomic super-computer** whose storage component (*DNA*) is operated on (for constructive, comparative and regulatory purposes) by *RNA* and protein. Whereas man-made computers work to base-2 (0 and 1) bio-computing works to base-4 (the nucleotides C, G, A and T). It also works in four dimensions.

Its storage string of *DNA* is, like any line, one-dimensional.

The second dimension, where parts of the string affect each other directly or through the abovementioned *RNA* and protein proxies, is regulatory. The latter act as repressors, activators, transcription factors and so on.

Third dimensional computation involves genomic shape. Genes are not randomly scattered but clustered according to need. Even if not neighbours on a chromosome, cooperative genes are collected to proximity by the way embedded code causes their relevant *DNA* to fold.

In the fourth dimension, time, nuclear conformations may change according to a stage of an organism's development. Epigenetic factors are, we'll see, also involved. Mobile elements called transposons and retrotransposons may also be involved in the process. The genome you end up with will not, at least in sequence or in sections used, be the one you started from.

Such complexity amounts to a teleological symphony. The bio-super-computer is a cosmic wonder. Only self-deception, a symptom of severe philosophical malaise, could believe any fully integrated, multi-layer control panel ever built itself for no reason by chance - even chance constrained by circumstance! **One would, therefore, predict research will more and more reveal signs of bio-logic to the point that, in 'live' computing, such complex permutations, integrated combinations and hierarchical sets of regulation will further squeeze then nullify the notion that celled systems ever cropped up accidentally, that is, evolved.**

A computer is a mind machine. *On this basis it is established that the cybernetic operations of a cell, an object as thoroughly material as a computer, superbly meet the criteria that pass it as a mind machine.* **A cell is a mind machine. Its instruction manual is a program written up as 'genome'; and its systems hold semantic meaning as modules or entities of bio-form like, for example, you.**

If this is right then naturalism's explanation badly needs to twist interpretation of the facts - for mind read randomness, for information chance. Henceforth we'll check the evidence against this warp.

Conceptual Biology

Is archetype a dangerous idea? For materialism it is fatal.

Bottom-up thus trenchantly denies it while increasingly the latest bio-science points 'upstream' towards a chanceless source. This source is, from a *top-down* point of view, derived from software called an archetype.

Software always issues from the purposes of prior intelligence; and archetype (or 'typical mnemone') **may be defined as a conceptual template or, in dynamic terms, a program.**

How, though, can frameworks that are 'logical', 'goal-oriented', 'reasonable' or 'purposeful' be implanted in biology? You may remember (from Chapters 11 and 12) *energetic* and *informative causation*. The former pushes you from past to future. **The latter is goal-oriented causation; and goals are in the future pulling you their way. They pull you *from* the future; they are metaphysical attractors, guides that govern your behaviour as they lead you through the world.** Of archetypal structure biology is generally concerned with instinct and the morphogene.[142] Morphogene means a specific, developmental program in universal mind, a metaphysical but also morphological attractor. Attractors pull things forward the way that, in response to pressing circumstance, they have to go. Push and pull cooperate. **Energetic and**

---

[142] Chapter 13: Typical Mnemone and *H. archetypalis*; Glossary: mnemone; Index; also *SAS* Chapter 16.

informative causation, running anti-parallel, are the way the world proceeds; and, in the context of biology, primary archetypes pull while secondary genes, body and its physical environment all push.

The basis of a conceptual approach to biology *is* archetype. It should be crystal clear by now that an unconscious archetype is, like instinct or memory, real but not physical. Indeed it *is* a memory, a natural record; it is a sub-conscious mental, metaphysical construct, an idea or rationale physically realised in a variety of similar forms. Call such template biological (since it's expressed as bodies) or (since it's in mind and includes instinct) psycho-biological the same.

For the information theorist 'archetype' is an idea realised as a main, core routine (or type of organism). Around this core are linked sub-routines - some universal, others more specific - which are appropriately tailored to integrate with each other under the control of any particular main routine. These main routines and sub-routines serve as modules; and they are recognised as homologies. Such a central *top-down* concept is worth rephrasing. **Various tailored sub-routines are called from a Main Routine.** Suites of such modules combine as permutations around which different bodies, each under the coordination of its own Archetypal Master Routine, are expressed.

In the *top-down*, hierarchical view of Natural Dialectic, tiered sets of modules compose, at top level, the range of life-forms and, lower down, their various coded and coherent parts. Again, the clearest example of this is demonstrated by the way that biological development is controlled. In such a scheme distribution of modules across the living world is, in varying degree, mosaic. For example, one that is critical (like respiration or nuclear operations) may appear, appropriately attuned, in every organism. Other sub-routines may occur in scattered and sometimes unexpectedly different locations (as for example, haemoglobin in humans and various kinds of plant). While evolutionists term such apparently unconnected, mosaic appearance 'convergence',[143] from the perspective of archetypal computation it is called 'modular programming'.

Molecular biology's great strides are also heading straight towards the notion of an archetype. A similarity of genetic instructions is found to extend across a great variety of forms. For example, a high-level developmental complex of modules (*see* Glossary: 'homeotic gene') may code for outline body-plan in very different organisms such as human, duck or fly. Or it may call subroutines for the construction of completely different kinds of eye - a normal gene from a mouse can replace its mutant counterpart in the fly; now an eye, a fly's eye not a mouse's, is produced. In other

---

[143] *SAS* Glossary and Index: convergent evolution.

words, such genes act as 'go-to' switches in an archetypal program of development.

The power of shape-making is equally evident in the operation of hierarchically integrated circuits of developmental regulatory networks.[144] Experiments as well as logic (for example, the engineering principle of constraints) indicate the impossibility of any kind of mutation gradually creating the tight-knit nucleic acid and protein agencies of such computer-like functionality. In fact, non-destructive mutations for latterly-expressed, minor, micro-evolutionary variations may occur but never for early, top-level genetic expression affecting body-plan. Macro-evolution of this sort is out.[145] Furthermore, the newly-discovered, fully-integrated layer of complexity, epigenetic programming, is observed to play a major role in morphogenesis.[146] Its informative cooperation still more radically contradicts the historical hypothesis of gradual, randomly-generated adaptive 'solutions-by-mutations' to ecological challenges. *At this point neo-Darwinian explanation reduces, simply in reality, to wilful materialism, a faith of naturalistic wistfulness and a misleading, occasionally polemicised mythology.*

And you rejected, philosophically out-of-hand, conceptual archetype? Variations-on-thematic-archetype are wrung, by breeders and mutation, on the forms of dog, horse and many other organisms. Yet the direction of natural selection, though at times supporting trivial differentiation, can't *create* a metabolic pathway, tissue, organ or a body-plan. It operates the wrong way - (↓) down towards *loss* of alleles and genetic information. Variants are, as even pedigrees may show, mainly due to bred-out loss of information and not evolution's necessary gain; then death always freezes up a limit to plasticity. **In fact, by artificially and fiercely driving such selection to its breaking point you'd soon empirically discover whether nature could snap barriers of type and innovate. There is no evidence, not even nascent, rudimentary evidence, it can**.[147] By contrast deliberate not random changes rung upon the archetypal sub-routines is how Natural Dialectic would interpret facts. It would informatively explain convergence, co-evolution or mosaic evolution as the engineering application of a theme to local detail, of a principle routine to different practices. *It would be found no accident that a mosaic conservation of both form and function pervaded somehow, somewhere, every genetic/ phenotypic form of life. These general*

---

[144] *SAS* Chapter 25: first three sections.

[145] natural selection and mutation are inadequate mechanisms by which to realise the transformist hypothesis (see *SAS* Chapters 22 and 23); modern genetics and molecular biology also increasingly militate against it; see Science 210 (4472): ps. 883-887, 1980.

[146] epigeny see Glossary; also *SAS* Chapter 21 nad *A&E* Chapter 8 esp. link 107.

[147] *SAS* Chapter 22: The Origin of Species; *fig.* 22.1 Plasticity; and Chapter 23: Adaptive Potential.

routines and sub-routines - locally reflected by material genes, molecular conformities and thence the larger structures that entirely constitute biology - would reside in immaterial mind.

In summary **top-down, hierarchical perspective rediscovers a primary bio-structural agent - archetype. Round this 'axis' secondary variation (due to adaptive potential,[148] sexual recombination and, to a lesser extent, genetic mutation) continually occurs.** *In other words, so-called micro-evolution is a secondary process which, far from progressing to macro-evolution, is dependent on the expression of a primary agent, conceptual archetype.* **Such a view, with concept preceding material chance, is logical, reasonable and data-compliant.** Its order reflects what we find everywhere. <u>However, because it includes an 'unnatural', immaterial element - intrinsic information - it implies the antithesis of a naturalistic, evolutionary account of origins.</u> For this reason the strength of the archetypal against the weakness of the evolutionary idea will now be briefly scrutinised.

Darwin: Half Right, Wholly Wrong?

Is not Darwinism mostly faith in the unseen? It is not Darwin's facts but his hints, suggestions, interpretations and extrapolations one might take to task. In so doing we'll check examples but, remember, this abbreviated version has lots more lined up behind.

Non-Darwinian seed of a Darwinian tree of life must have evolved from barren pools or clays; from mud or (either warm or cool) saline solution. Could this be physically possible?

Perspectives on Three Central Tenets of neo-Darwinism - a Tabulation			
✓ true ✗ false		Bottom-up	Top-down
①	Abiogenesis	✓	✗
②	Variation (microevolution) by mutation and natural selection	✓	✓
③	Transformism (macroevolution) by mutation, natural selection or any other means	✓	✗
Aren't half-truths the hardest ones to disentangle; and ones with greatest tendency to lead astray? **Indeed, neo-**			

---

[148] *SAS* Chapter 23: Super-codes and Adaptive Potential.

> **Darwinian evolution (as opposed to biological variation) may be viewed in the light of a logical fallacy, a trick called equivocation.** *This semantic trick is to conflate two entirely different matters - firstly, variation on existing features (tenet 2) and, secondly, tenet 3's <u>addition</u> of complex, highly informed fresh features (such as coherent organs, systems and so on) in the first place. You might even try to slip in tenet 1!*
>
> Over the next few pages we shall ask:
>
> 1. if bio-monomers (such as amino acids, nucleotides, sugars, lipids and so on) could spontaneously form in a single place in sufficient quantities to build a cell.
> 2. whether interconnected bio-polymers (such as proteins and nucleic acids) could form from these in water.
> 3. whether the integrated, codified metabolic pathways that a cell requires could appear in operational readiness from scratch.
> 4. whether a cell, operating in homeostatic, steady-state disequilibrium, could be constructed by chance in a continually changing external environment.
> 5. even if it were, could such dead-ringer of a living organism (an extremely well-informed, complex and huge conglomerate of atoms) be 'raised from the dead'? Could the pre-life corpse be rapidly primed to a dynamic, steady-state condition before chance decomposition or fatal degradation of any part occurred? Could life accidentally 'go live'?
>
> The holistic answer is, of course, no. On what evidence is naturalistic speculation, strongly theory-driven, based?
>
> Next the mechanical possibility of macro-evolution is questioned. On the ground, there is practically no evidence for this crucial extrapolation from variation or so-called micro-evolution.[149] Surely the evolutionary scenario, endless possible plasticity, is not composed by only hope's determination and a flair for rational story-telling?
>
> **fig. 15.2**

Which inference, yes or no, do the facts (most of which Victorian science did not know) support? If they support a positive then Lady Luck has won the toss. If they support a negative then 'life-from-non-life' flips to 'modern alchemy'. *An axe is laid upon phylogeny; the root is sliced; Darwinian seedlings never sprang to sprout a tree.* **No seed, no tree.**

---

[149] *SAS* Chapter 22; *A&E* Chapters 5, 12-15, 17.

Chemical Evolution?

For a start, the issue is not one of religion or opinion but science and logic.

One explanation for our beginning is that we were deliberately created. The other, which is the neo-Darwinian theory of evolution, is that we were not. The latter starts with a process called chemical evolution.[150] This phrase implies that lifeless chemicals, having long resisted the degrading force of entropy, 'co-evolved' in a sub-microscopic space whence they could 'self-construct' the primary unit of life, a reproductive cell. It guesses the generation, perhaps gradually over a long period of time, of life from non-living components by physical means alone. Hail, Bacterial Patriarch! The theoretically life-imparting process[151] is integrally part of, strictly not the same as, Darwin's consequent evolution. After all, it casts no role for natural selection, variation or mutation; Darwin merely hinted, hopefully, that once upon a warm pond....

This book is an abbreviation. Therefore, let famous synthetic chemist James Tour abbreviate in an essay (2016) entitled 'Animadversions of a Synthetic Chemist': "Those who think scientists understand the issues of prebiotic chemistry are wholly misinformed".

E cellula omnis cellula. Only from a parent cell does daughter come. **No exception has ever been found to this rule so that it is called The Law of Biogenesis**. In 1860, a year after the publication of the Origin of Species, creationist Louis Pasteur decisively debunked the medieval notion of abiogenesis (then called spontaneous generation) by demonstrating that broth in sterile flasks did not spoil. This experimental discovery, which underwrites the use of sterile equipment in medicine, therefore showed that life does not spontaneously arise. It confirmed the Law of Biogenesis. *Nor in modern times does any process of abiogenesis, whether quick or slow, man-copied or natural occur.* Materialism is forced into the paradoxical and self-contradictory explanation that information comes about by chance. A universe that could not make a cup of tea (let alone with milk and sugar) in a billion years is supposed to have gradually produced the efficient, codified development of a tea-maker! This unproven creation story conforms to the philosophical requirements of naturalism and is, for this reason, supported by research grants and currently popular.

What's the Problem?

School taught you Darwinism works. The idea (random mutation acted on by natural selection) is simple. You received the *AVB* (Authorised Version of Biology) but nowadays its gloss is wearing very thin.

---

[150] *SAS* Chapters 20-21 and *A&E* Chapters 4 and 8-11 for an accumulation of reasons why chemical evolution is a complete non-starter, a modern form of alchemy born of materialistic mind-set alone.

[151] also called abiogenesis, biopoesis or prebiosis.

Can selection of mutations generate sufficient bio-novelty?[152]

How can the effect of natural selection be precisely measured? Can the effect of mutation as regards the innovation of integral and integrated organelle, cell-type, organ system or body-plan be quantified at all? Or can you exactly calibrate when variation stops and macro-evolutionary transformism starts? If you cannot then evolutionary theory reduces to the merely anecdotal.

Mutational 'creativity' needs synchronously include the generation of a digital information system (grammar, molecular agents and operating systems) from scratch; the original generation of epigenetic co-systems and of hierarchical, computer-like routines of government whose tightly integrated circuits every cell displays; rapid innovation of new form (e.g. Cambrian body-plans[153]); the origin of sex and developmental algorithms; the morphogenetic congruence of hierarchical layers of construction ranging from molecule through molecular coordination, organelle, cell, tissue, organ and system to target body-form; and the evocation of irreducibly-complex mechanisms at all these levels simultaneously (if things are to work).

The problem is, of course, not the origin of species but of concentrated information; and, in such magisterial dock, evolutionary theory repeatedly breaks down.

A systems engineer at ARM or INTEL would, if you announced his complex chips arose by chance, evince surprise; no less if you contended that the mode of his construction could, equally, have been by gradual accumulation of improvements mindlessly. Because denying information's source in mind denies its causal agency; and as we know, but Darwin didn't, symbolic information is the basis of biology. Though the discipline's dramatically progressed still evolutionary theorists with

---

[152] Mathematical challenges to Darwinism have long queried the enormous number of beneficial mutations needed not only to create biological functionality (a factor erroneously taken for granted by evolutionists) but also to improve different parts of interlocking subroutines simultaneously. In fact anything produced by a random process is far more likely to be destroyed by that same process than be improved by further sets of 'lucky change'. Perhaps the first powerful mathematical challenges to the neo-Darwinian interpretation of evolution were issued by the Wistar Institute in 1967. Logician Ludwig Wittgenstein was also very skeptical of the theory; so was mathematician Paul-Marcel Schützenberger. Indeed, increasingly powerful resistance by scientists and mathematicians to the theory has developed since the seed sowed in 1932 by Ambrose Fleming, inventor of the thermionic valve, effectively the device giving birth to modern electronics and the IT industry. Although the Darwinian account is replayed as repetitiously as a popular record, there is no shortage of highly intelligent rebuttal.

[153] *SAS* Index: fossils

naturalistic goggles firmly pressed deny such agency. They seek inference and interpretation whose 'best explanation' categorically rejects an immaterial element - although we know that programs, books and speech are physical expressions always traceable to such an element.

'I remember at an early period of my own life showing to a man of high reputation as a teacher some matters which I happened to have observed. And I was very much struck and grieved to find that, while all the facts lay equally clear before him, only those which squared with his previous theories seemed to affect his organs of vision.'

Lord Joseph Lister (1827-1912)

Evolution is agendum-driven.[154] *The problem is profound because the whole edifice of materialism demands primordial seed from which its 'family tree of life' can then branch up or, if you like, can commonly descend.*[155] If, therefore, the hypothetical chemical evolution of a proto-cell *is* no more than sophisticated animism or an alchemical myth, then what of transforming it by so-called evolution into plants, trees and all the panoply of life on earth? Subjecting the interpretations of neo-Darwinism's modern synthesis to an interrogation, specifically we mark:

1. whether mutations and natural selection are sufficient mechanisms to generate, at most, more than variation-on-theme, called micro-evolution,[156] that we observe continually occurring.
2. if they are not, and since no other mechanism has been found (with mind or archetypal memory disqualified), what might generate the large influx of highly organised information needed to innovate a cooperative biological pathway, process, organ, system or organism? Neither innovation nor transformation has, on the large-scale required, ever been observed; upon close inspection not even nascent systems are found.[157]
3. organisms involve complex editors[158] to reduce rates of mutation and preserve the integrity of code; their effect actively minimises any chance of evolution. Moreover informative entropy also scrambles and deletes 'progress'; everywhere, both now and by extinction in the past, loss or degradation of genetic information is the norm. How, upstream against the torrents of this natural, anti-evolutionary deluge,[159] could complex and

---

[154] see Glossary: evolution; also *AMA?* Chapters 1-6.
[155] *SAS* Chapters 0, 22: The Origin of Type and 23: Evolution in Action?
[156] *SAS* Chapters 5, 6 and 19-25; *A&E* Chapters 4 and 5; and *AMA?* Chapter 22.
[157] *SAS* Chapter 23: *passim*.
[158] *SAS* Chapter 23: Non-protein-coding *DNA* and Hierarchical Language, Super-codes and Adaptive Potential.
[159] *SAS* Chapter 23: Entropy of Information.

correctly-sequenced order-building due to accidents occur?[160]
4. fossil evidence is used selectively against a chosen theoretical criterion (popularly, evolution); similarities in fossils may or may not be due to ancestry; but, in principle, fossils cannot establish unseen lines of ancestral relationship; 'progressive sequences' thus constructed are imaginative, *ad hoc* illusions. Therefore, how can the fossil record substantiate an idea of macro-evolution?[161]
5. how sex, development or any other agent of reproduction and development might have 'emerged' by chance.[162]

Attempts to constrain systems innovation within an evolutionary framework has also dreamed less than forthright accommodations.[163] For example, atheism's 'useful idiots' assert the existence of laws or processes for which there is no evidence. *Deistic evolutionists* (like Erasmus Darwin and his grandson Charles) allow an impersonal, 'externalised' kind of creator, the miraculous (metaphysical?) creation of souls and of lawfully-behaving energy. Post-creation, however, such creator neither creates nor plays any interactive part in life; materialism's Darwinism makes life up at random. *Theistic evolutionists* don't believe (correctly) that undirected chemistry can single-handedly create specific information, generate a functional system or, at length, write up humanity. Yet such theist is mind-locked into a Darwinian 'progression'. Wind, rain and natural forces *never* codify and, since there exists no Law of Non-conscious Innovation and Integration, no Principle of Material Evolution then, accepting a deistic start, the fellow effectively embraces a notion of *intangibly divine tinkering*. However, all the variation that we *scientifically* observe arises from sexual recombination, adaptive potential (see Glossary) and genetic accidents; guessed-at interventions (say, sets of purposeful mutations or hopeful monsters born radically deviant from parental forms) are, emphatically, unobserved. The Intentional Tinker is a mischievously absent imp!

No need, on the holistic hand, for fairies. A prediction is that bio-information systems will, for every cell, prove at least as rapid, precise and complex in operation as a highly powered and, of course, informed computer. ***At the same time, forthright answers to the questions may well indicate that neo-Darwinism isn't fit and doesn't work; that promissory faith in future answers is, if not a disingenuous ploy, still insufficient to delay for long the theory's demise; and that, in the light of modern science, evolution's vital alchemy demands a radical revision.***

---

[160] *SAS* Chapter 22: Origin of Species, Galapagos and All That; The Tree of Life, Homology and Origin of Type.
[161] *SAS* Chapter 22: Types of Fossil.
[162] *SAS* Chapters 24: Sex and 25: Development; *A&E*: Chapter 7.
[163] *SAS* Chapter 25: Theories of Accommodation.

Chapter 16: Truth, Appearance and Reality

Truth, Appearance and Reality [164]

	tam/ raj	Sat	
	lesser/ existential being	Being	
	lesser truth	Truth	
	lesser reality/ appearance	Reality	
↓	tam	raj	↑
	objective	subjective	
	matter	mind	
	from Centre/ towards triviality	towards Centre	
	lesser truth	greater truth	

Bottom-up, reality is physical. Materialism's primary axiom is set. The rest, excepting maths and scientific fact, is simply story-telling. Thus everything, including consciousness, is construed as a 'material phenomenon'. Such is, although it might be absolutely wrong, materialism's promissory faith. What you can slap is obviously real. Yet, even here, Niels Bohr claimed that we accept as real what's not. Quanta substantiate the large-scale universe which, on such terms, is an appearance. Matter is, in this sense, really energetic patterns and grosser views are less than fundamental. Part of physical reality's invisible. To this extent its truth's incomprehensible to common sense. How does grey matter grasp this fact?

No doubt, brains are lenses. They are sophisticated interlinks between mind, body and the latter's universe. As such they cut to a local portion of reality. They are constructed, you may claim, with such capacity and in the way that an embodied human's *meant* to know the world - but is 'mind-meat' alone enough to sense and understand? From a material perspective perception is, in fact, nervous activity. In other words, just find a 'neural correlate' (the pattern of the way synapses form and fire when using different parts of brain) to find perception's underlying truth. Where electric signals storm like drops of rain, there at grey stuff's neural base you'll find mind's rainbow crock of subjectivity. Matter is the truth; mind and metaphysic are appearance; physic is self-evident reality.

If, however, mind is *not* meat then, *top-down*, perspective flips. Every organism knows its world (that is, its reality) from inside gross and subtle frames. Gross is the non-conscious, physical body; subtle is instinct. Many organisms, including humans, are also subtly framed by personal memories. These three filters colour a local conscious experience; they frame interests and purposes within a sensory, creative

---

[164] also *SAS* Chapter 4.

or contemplative focus of attention; and they limit sense of what is going on and thus restrict our truth to personal experience. The apparently inescapable combination, one part physical and three metaphysical, is commonly identified as 'me'.

The metaphysicals are bundled up as 'mind'.[165] After all, an atom isn't conscious, mindful or informative. Why, therefore, should a thousand or a billion be? Congregation of non-consciousness doesn't just 'become alive'. Oblivion is every atom's everlasting state. Is consciousness composed, therefore, of oblivious matter or an immaterial element? Though not permitted by the rules of physics its unproven metaphysic is each person's base reality. Not physically provable may not mean non-existent; don't immaterial thoughts themselves create our science and mathematical description? **Certainly, if there's reality *outside* our consciousness we'll never ever know it!** Nor is anyone aware of undiscovered facts. Consciousness is all we have to know with. It is life. Experience, feeling, knowledge, logic, math and meaning utterly depend on it. Are these unreal?

Holism, you are well aware by now, does not admit that mind's a function of a complex form of matter and, in essence, physical. In this view all things *aren't* equal. *The illusion of illusions is that consciousness is an illusion*! What, except materialism, might indulge such fantasy? **Consciousness and matter are elemental components of cosmic duality; information and energy are distinct, complementary aspects of existence.** This line of reason leads to a hierarchical or vertically-graded structure within which truth, appearance and reality are evaluated. **If Absolute Truth manifests a hierarchy of appearances, if from Substantial Reality emerge relative illusions and, therefore, things are not equally real, what is the criterion by which one is judged more or less real than another?**

It is as if, down the Cosmic Mountain's slope, different levels of perspective are obtained. Perspective changes as you rise or fall. Does love or understanding give the answers ignorance or hate purvey? Such a Theory of Relativity with respect to Truth, Appearance and Reality is actually neither new nor strange. Buddhists, for example, are taught appearances can be deceptive; the sensible world of trees and people and teacups is a partial illusion; material circumstance is relatively unreal. Science now knows that, from the non-sensible perspective of quantum physics, this is correct.

Are not the Buddhist notion of perpetual flux and the modern theory of kinetic matter in essence similar? In the latter, constant motion and vibration of bulk matter's make-up - quantum elements and atoms - explain such basic features of our world as temperature, diffusion and phase change. In the former, explanation is extended to include a medium

---

[165] see Chapter 13.

of information - mind that also oscillates, radiates and can't keep still at all. No permanence. Vibrant motion drives appearance; it constitutes existence. Science well describes its protean object, energy. But whence the latter's source; what in reality substantiates the ripples of existence? And what, when physical and psychological illusions have completely cleared, is a Buddhist's Absolute Reality?

	relative illusion	*Truth*
	less right	*Right*
	critical comparison	*Criterion*
	lower qualities/ lesser values	*Quality/ Value*
	shades	*Illumination*
	degrees of incomprehension	*Clarity/ Clear Mind*
	lower principalities	*Principal*
	death/ life	*Pure Life*
↓	*lesser truth*	*greater truth* ↑
	blinder mind	*clearer mind*
	away from 'right mind'	*towards 'right mind'*
	lower qualities/ separators	*higher qualities/ unifiers*
	quantifiable objects/ events	*experience*
	non-conscious forms (matter)	*conscious forms (mind)*
	individual/ local contexts	*principles/ symmetries*
	less/ least real	*more real*
	towards darkness/ oblivion	*towards light/ understanding*

Appearances can be deceptive. If cosmos is projected it appears from source. Its appearance is effect not cause, lesser truth not basic. On this basis physic's cosmos isn't Real. Matter isn't Real. If principles inform behaviours then informative causation governs energetic patterns of effect. This sort of world-scale means a starry universe of matter isn't True - it's an appearance, only relatively true, projected from its hidden yet informant metaphysic, Truth.

Is there such Ultimate Reality? Belief, at least, depends upon criteria or, absolutely, your Criterion of Truth. If Top Pole is positioned as Creation's Source then you might claim that it's most real. Transcendent First Cause throws a hierarchy of appearances whose shadows deepen with their distance from its Light.

<u>*In short, there is a sliding scale of truths but also, at the top, a Truth of truths*</u>.

In terms of information a qualitative hierarchy drops from accurate and important through to incoherent, valueless or even malevolent. At the base of subjective worth a special case, non-conscious matter, is worthless. Ashes to ashes, dust to dust; but it forms the basis of a vast, magnificent, dead universe.

In terms of energy potential is expressed through action to exhaustion; it is constrained from possibility to single actuality, the end-result.

The greater reality is, as with Bohr's quanta, vested in the invisible origin of any particular expression. At mental level this source/potential is called *Logos*; and at material level 'potential matter of the archetypes'. If potential precedes action then we have a sliding rule - metaphysic unto physic - for priority, importance, truth and reality. Relative truths are measured against Absolution; such Top, Exemplary Criterion is embodied as a Perfect Saint.

Of course, such scale[166] is very different from the popular materialistic view of cosmos but it strikes a deep chord with monastic, mystic and non-materialistic frames of reference as variously expressed throughout all human history. It would seem its logic is superior to flat-earth, one-dimensional materialism's standards and it certainly implies Darwinian evolution as a creation-story is not able to ring more than fractionally true.

In summation, scientists keep finding what nobody knew before. Such enterprise, although its truth is thereby relative, still seeks material absolution - perhaps in the eventual proof of a Grand Unified Theory or Theory of Everything. What, however, if its thrust ignored a whole dimension? Or denied that immaterial elements of psychology are actually immaterial - especially if, in absolution, Truth of truths *were* of the immaterial, metaphysical dimension? And what if, from that Absolute Perspective, all mental and material phenomena involved a scale of changing, relative appearances. From Absolution proceeds relativity. Thus, could materialism's atheism ever claim, even in principle or promissory future proof, to know it all? And if that Truth were Living then it would forever miss The Point.

Two Value Systems

We have glimpsed perspective that is, at first glance to a sensible, down-to-earth way of thinking, strange. Are you now ready for a fresh dose of dilemma - this time of another sort? Value, meaning and significance arise from mapping cosmos by experience; each new experience within this context is imbued with its own quality of truth and truth must, more or less, bear worth. Moreover, moods and machinations make, at different levels and with differing objectives, various value judgments; such choices are infused with a person's overall perspective called a world-view - *top-down* or *bottom-up*.

Philosophy today is ambivalent. Are mind's illusions born of gene mutations? Are they only nerves or not? Some cannot quite believe that everything is physical. Others, hard no-hopers, can't accept the metaphysical at all. Their saturated scientism demonstrates monopolistic tendency. Such

---

[166] A scale of higher or lower being with more or less reality was hinted at in Chapters 8 and 9; see also Index and *figs*. 11.3 and 13.3.

quantity/ quality	*Quality*
objective/ subjective	*Subjective*
range of values	*Value*
lesser truths	*Truth*
↓ *lesser truth*	*greater truth* ↑
external	*internal*
objective thing	*subjective sentience*
physical context/ bodily self	*mental context/ egotistic self*
quantity/ aggregate	*quality/ meaning*
value of things	*value in mind*
numerical/ market/ bodily value	*motivating/ emotional value*
less important	*more important*
utility	*beauty*
using/ abusing/ careless	*caring*

authority would dictate how a human must explain his cosmic state. Many strands are complicated in its uniform; each thread is spun from physics' loom; and all are woven into 'there-is-only-matter' patterns but must still, grudgingly, acknowledge intimation of some immateriality in the form of information, will-power, creativity and, due to choice, discrimination of morality. Perhaps you'd prefer, since you're a ghost in the machine, that there was no psychology - just neuroscience, electronic circuitry and the biology of nerve. Then it's easy - evolution made the whole thing up.

That, then, is who you are! Amoral molecules cannot beget morality. Since man must be the product of rude chemistry his body's nature is defined by genes and their surrounding circumstance; and mind's experience is nurtured by a nervous network of relationships. Thus are goals atomistically conceived; and by their yardstick what is judged success or failure must accrue. Critically, what is the value calculated? What is, when it comes to goals, a man's most valuable choice? Comfortable survival turns up trumps! No doubt, therefore, that politics and economics rule a creature's day.

'Quantity' of life-style is the objective business of economy - resources, wealth and body-care. And scientific values are non-ethically composed of numbers and utility.

'Quality' of life-style, on the other hand, involves subjective business - interests, relationships, aesthetics and happiness all round. Such socio-economic linkage needs firm leadership. A population is a clamorous 'family' composed of selfish and thence often spiteful members. It needs more or less a strict, *external* government in order to promote group balance and to mould conflicting instincts, interests and behaviours into as stable yet dynamic a pattern as can be. Equilibration and morality combine to exercise a healthy body politic. The quality of equanimity is, however, optimally based not on external but *internal* government. Such self-imposition stems from moral principles by which a person lives; and these in turn are coloured by the pillar of his faith. So civilised! In an ideal city of the bless'd and blessing there would exist no needy, selfish sinners. Nor would vice hang expensive millstones round the neck of virtue. *Costless, painless law and order derives from principles - invisible, immaterial, unscientific principles.* Yet in our time it sometimes seems that neither clever socio-biologists with selfish genes nor cunning crooks respect the timeless root of crimelessness.

Top-down vision is consolidated round the Central Axis of Enlightenment. Its holistic value system attends Nature or (if you insist that nature's only physical) Super-Nature. Its Highest Value, the First Principle from which all others flow, is equated with the Nature of Innermost Self, The Highest Good or Apical Experience. Such Experience is extolled. It is symbolised at the sacred heart of all world faiths and in their personal devotions. This central residence would, dialectically, deposit any cult of matter (including humanistic scientism) at the periphery of truth; it would define a *bottom-up* perspective as, according to its concentration of materialism, a relatively eccentric misconception. This eccentricity occurs because materialism stands on fundamental Lack of Reason. Must nature's automatic patterns have their origin in chance when chance, by definition, lacks all reason? If so, materialism's faith in lack of reason is irrational and atheism's hopeless instinct doesn't sum to immaterial faith at all.

Do you recall inversion?[167] One could turn the tables and flip Natural Dialectic wholly on its head. Upside-down. Where head, now toe. Materialism's tabulation diametrically twists the cosmos round.

relative truths/ falsities	*Axis/ Basic Truth*
immaterial figments	*Matter*
soul/ mind	*Mindlessness*
degrees of awareness/ life	*Non-Conscious Oblivion*
ghosts in the machine	*Blind Automaticity*
delusion of morality	*Amorality*
subjective metaphysic	*Objective Physic*

Such tabulation represents holistic anti-truth. Its 'anti-stack' systematically twists, reverses and thus overthrows holism's rationale. Nor does it, lacking an accepted metaphysical dimension, countenance cosmic fundamentals, immaterial mind or conscio-material gradient of reality. Matter and material vectors are its everything. Therefore, it denies holism wholly; holism is the enemy of atheistic state. For this reason such atheistic mode of thinking has, from a *top-down perspective*, been identified as The Illusion. Its confusion casts, because its snare serves as a devil's net, a

---

[167] the flip of one extreme of character to its opposite; see *fig.* 10.3, Chapter 10, Glossary and Index: anti-parallels/ dialectical factors/ inversion.

diabolically flattened, toxic form of Natural Dialectic. You could build up anti-stacks and work this out yourself.

Rights and Wrongs

From a value system flow one's reasons. These calculate, accordingly, what's right or wrong. Whether it's a thought or action, 'right' decision sails you home and dry.

relative shadow	*Light*
range of rightness	*Rightness*
scale of inferior reasons	*Reason*
lesser aims	*Ideal*
↓ *negative*	*positive* ↑
anti-principle	*principle*
downward/ outward	*upward/ inward*
depression	*uplift*
from Reason	*towards Rightness*
false/ wrong	*right/ true*
passion	*patience*
obsession	*detachment*
pain	*peace*
sin/ vice	*virtue/ righteousness*
towards death	*towards life*

Do right and wrong,[168] like ought and should, depend on metaphysical criteria? Who or what decides what's good or bad? Since every brain is different, might we materialistically deduce no absolute criterion exists? Neural products known as preference, logic and morality should vary in each one. Moral relativity is this game's name. If ethics and aesthetics ooze from nerves in brain then codes are cerebrated as the molecules dictate; and since brain configurations change our values vary with them! Why, anyway, should anybody, much less Him or It, tell me what to do? In this view 'personal or social opinion' blurs with absolute morality. Why should enslavement, murder, rape or theft be 'wrong' if that's the way I naturally think? Survival justifies self-seeking plans. You don't jail cats for killing mice and we're just animals as well. At least, this is the sort of creed that evolutionary tenets justify.

The problem is that such society of relative morality (and thence, politically, its laws of state) is built on shifting sands. <u>Without an absolute reference-point trivial issues are confused with major</u>

---

[168] the problem of evil is addressed in *SAS* Chapter 26; see also the previous section's diabolical dialectic.

principles. Each man judges the excellence or otherwise of a given behaviour differently; such a secular society drops into an ethical morass of fashionable 'political correctness', changeable, relative moralities and even, at nadir, moral meltdown.

From Science to Conscience

Oblivious maelstrom is valueless, meaningless and (*pace* atheism) cannot generate a code of any kind. Rational, legal or moral codes don't spring from witlessness. Thus study of material is not the same as study of the immaterial. Laboratories are not the place for moral seminars. *Indeed, scientific progress is irrelevant to moral progress.* Nor does the latter's personal kind of evolution change with history. Thus scientific materialism may exclude reference to moral struggle and human social context or, at best, objectively describe these in animal (socio-biological), humanistic or politically utilitarian terms. Such reason does not render subjectivity unreal. *There are different truths and therefore ways of seeking them. Do many ways up to its Apex mean a mountain lacks a Peak?* Why is it rational to deny the Universal Height is immaterial - unless a metamorphosis of reason called materialism has forgotten reason is itself a form of metaphysic?

Mind evolves by lowering its own entropy of information viz. increasing comprehension. No doubt, clever intellect pursues its variously important disciplines but wisdom comprehends by depth of empathy. Either's orderly simplicity derives from principle; and the more embracing a principle the more details are easily subsumed under it. Principles are concentrated power. They are the rule of natural and man-made law. If, therefore, the Axis of Mount Universe represents the Central Principal of Principles then the ascent of wise man will involve his mountaineering towards this Most Essential Peak. **Enlightenment is, naturally and automatically, Morality.**

Such morality's as natural as gravity. Its levity is unavoidable. Such immaterial but crucial datum is not gleaned from a laboratory bench. **Indeed, the fact is that all the emperors, generals, politicians, philosophers, philanthropists, artists *and* scientists in human history have not exercised as much influence as a few perfect mystics.** Waves of their transcendent experience resonate, as they seem always to have done, in the heart of mankind. Sages such as Buddha, Christ or Nanak (in alphabetical order!) personify Ideals. How, therefore, might you realise their fantasy of phantoms, an imagination made of neural networks and patterns produced, in the last material analysis, by randomness?!

Is There an Absolute Morality?

If the most powerful curative for personal, social and ecological ills is the right psychological perspective then the question becomes 'What is most right? How do I recognise it and obtain its Rightness?' Such questions have always absorbed mankind; they encapsulate his quest for Truth.

Bottom-up, reason and not Reason wins. A reasonable, atheistic humanist may well, in bare-faced self-contradiction, reject Darwinian 'morals' (nature-red-in-tooth-and-claw, war-like survival-of-the-fittest and so on). In reality, there's no confusion since, in all bio-societies, non-Darwinian rules deliver the behaviours of peace. So he resolves the conflict by adoption of an immaterial code and thence behaves 'as if' some godless and yet god-like, altruistic creed prevailed. At the same time, with his finger on your pleasure spots, this clever humanist seems to discard morality's control. Who'd be fool enough to disagree? Comfort, if not hedonism, is 'the good life's' game. A siren voice excites an easy popularity. Just scratch my back and I'll scratch back - especially if you're 'family'. To hell with medieval sackcloth! Let us maximise on fun!

Top-down, reason is aligned with Reason. Conscience-in-practice starts, with an appreciation of the right ($\uparrow$) and wrong ($\downarrow$) vectors of behaviour, at the seat of mind - the eye-centre. It is here, where centaur-like humans leave behind the body's animal, that choice between a course of action and its consequence is weighed; and therefore here, where they can voluntarily transcend instinct and *think*, that moral rudiments are realised. From this point, X, their cosmological axis,[169] a first positive, purposive and therefore, obviously, voluntary step towards higher truth is made. *It involves, through meditation, the achievement of a highly coherent or concentrated focus of attention at the eye-centre.* From this point the goal, following an inward transport, is to reach the top of Mount Universe and thereby achieve identity with the Centre. **This identity is with the Principal of principles. It is called Enlightenment - in which** *top-down* **perspective is entirely obtained. Teacher, method, practice and completion in identity together constitute a science of the soul.**

In short, if the cosmos is stepwise in structure, then well-aligned philosophy and good, sound mind should reflect the fact. In other words, there will scale a gradient of qualities of mind and, logically, transcending existential relativity one might aspire to Highest Good, Noblest Truth or The Criterion. **Such achievement *is*, intrinsically, Essential, Absolute Morality.**[170]

Bodies differ, minds are not the same but, in essence,[171] souls are one. If the basis of creation is its Soul then cosmic peak is Union. The nature of Communion is empathy, sympathy, love. All lesser loves head, more or less, ($\uparrow$) towards this Love. What more to say? Your Lord is Love so "Love the Lord with all your heart and love your neighbour as yourself". **This approaches, automatically, Morality.**

Q.E.D.

---

[169] see *fig.* 7.4.
[170] *SAS* Chapter 26: Individual Association.
[171] Chapter 8.

Appendix 1

Natural Dialectic promotes systematic enquiry. Its scope, covering metaphysical and natural philosophies (the latter known as science), is encyclopaedic but seeks, above all, to reduce creation to simple principals and principles. So its columnar representation easily and orderly cuts to the heart of oriental elements as formulated in Taoist, Confucian, Hindu and Buddhist expressions of universal order. Its cosmic fundamentals (or basic trinity) also relate to the way Christianity explains mankind's experience of the universe. How?

As regards Chinese philosophy:

↓ *yin*	*Tao*	*yang* ↑
existential passive	*Essence*	*existential active*
down	*Balance*	*up*
increasing inertia	*Peace/ Poise*	*increasing energy*

Tao is an Essential Character whose Pivotal Way of Balance is at the heart of Taoist mystic philosophy. You strive for moderation and the middle way. Lao Tse is its father philosopher and the Tao Te Ching its book (although The Book of Changes reckons it as well).

↓ *yin*	*yang* ↑
phoenix	*dragon*
female qualities	*male vigour*
coolness	*heat*
mass/ earth	*energy/ sky*

It is said that, in 518 bc, Lao Tse welcomed Confucius (Kung Fu Tse) to Beijing. Mystic met pragmatist. Confucius drew the Way of Essential Taoism down to the existential world of polarity evident in society, economics and politics. In his Analects he formalised The Way of Harmony with respect to nature's complementary duo. Of this pair *Yin*, the (↓) female force, is symbolised by a phoenix; she signifies coolness, beauty and female qualities. *Yang*, on the other hand, is the energetic (↑) male factor symbolised by a dragon. Their *balanced complement* is extolled in a form of symmetry (where symmetry is an aspect of balance) called Feng Shui. The Confucian 'set' of priorities were formalised in hierarchical social structure and rituals which, once adopted by emperors, constituted the *top-down* canon by which Chinese government and family were organised. The code is enshrined deep in eastern psyche.

At the heart of The Golden Rule of Moderation is a morality of empathy. At the head of its hierarchical system of respectfulness The Father Emperor acts as middle-man between the Heavenly *Tao* and earthly population. Order rules, chaos doesn't. Randomness is exiled, disorder is repelled. In a wise society Confucian ideals and virtues permeate every family and individual. Such respect for *Tao*, emperor, his civil service, ancestors, elders and parents endures to this day. It embodies the Chinese state of mind.

↓ *tam*	*Sat*	*raj* ↑
materialisation	Brahman	dematerialisation
Brahma/ Saraswati	Vishnu/ Lakshmi	Siva/ Parvati
creator	Precursor	destroyer
down	Balance	up
disintegration	The One	unification

As Christ was a Jew the Buddha was a Hindu. As regards Hindu/ Buddhist philosophy the situation is slightly more complicated. Brahman is the formless reality, the ultimate source of creation or, in terms of Natural Dialectic, Essence of existence. And, where a human is capable of attaining divinity, for a Buddhist realisation of this 'ground-state' is called *Nirvana*. The Hindu sage knows such communion as *Samadhi*.

From source the cosmic grid is transformed down a scale of creation. Its power, likened to the vibrant harmony of song, has many names. Its Christian 'order' is called *Logos* or the Word of God. In Hindu terms the power of *Om* is dissipated from *Sat Nam* (True Name) down to the level/ waveband of psychosomatic *prana*. Metaphysical *prana* is, the psience suggests, finally reduced to base, non-conscious level. This weakest region is the physical projection called, with its particles and forces (such as light, electrons and mass-centred, contractive elements), our physical universe.

However, the popular Hindu version of order is nourished by way of a theistic trinity. These three, each representing a cosmic fundamental, are (*Sat*) Vishnu, aspect of Balance who sits central between Brahma (the *tam* ↓ materialising force) on the left and Siva (*raj* ↑ dematerialising stimulant) on the right. This pantheon is slightly complicated in as far as each such deity is in turn polarised; the complementary female expressions (or wives) are called, respectively, Lakshmi, Saraswati and Parvati.

↓ Son	Father	Logos ↑
outward/ overt	Central	inward/ intrinsic
externalisation	Source	internalisation
creation	The One	integration
body	Soul	mind
detachment		communication

The Christian trinity is a further, powerfully personalised expression of cosmic fundamentals. The players are, paradoxically, three-in-one. From Father issues the metaphysical loop of *Logos*, both (↓) creative and (↑) integrative; and, at this loop's physical extreme, He is reflected in the incarnation of His Son. The Son's inmost reality *is*, Logically, His Father's. The message is that, in communion with their own true nature, a son or daughter may be drawn back through *Logos* to Most Natural Source, that is, to The Father or Unsexed, Essential One. **The way to complete creation's loop is, although described in various ways and called by different names, essentially the same for all human beings.**

Glossary

A

adaptive potential: involves pre-programmed, super-coded switches and recombinant (transposable) refinements intrinsic in the genomic program of any particular biological type (*SAS* Chapter 23).

allele: a sister gene; you have two copies of life's book, one from mother and the other from father, so that each gene from father has a correlate 'allele' from mother and *vice versa*.

anti-entropy: *see* **negentropy**

archetype: basic plan, informative element; conceptual template; pattern in principle; instrument of fundamental 'note' or primordial shape; causative information in nature; 'law of form'; nature's script; Natural Dialectic's 'holographic' edge; the psychosomatic place where metaphysic and its physic meet; morphological attractor or field of influence in universal mind; the subconscious component of universal (natural) mind comprising archetypes; prototype-in-mind (maybe related to Platonic ideas, Aristotelian entelechies and/ or Jungian archetypes) whose potential matter is seen as hard a metaphysical reality as, say, particles are physical realities; like mind, being metaphysical, archetypes are physically non-existent; they are unobservable except by inference; materially formless and boundless, they are in this sense infinite; the fixed source of material projection, an unconscious archetype, is omnipresent and omnipotent; conscious Archetype (First Cause Psychological) is active (not a memory) and also omniscient; as all orderly processes involve a program, so the passive, natural preconditions for our material universe are stored in cosmic memory - simple in terms of inanimate physical 'law' (of particles and forces), complex in terms of animate structure/ function/ behaviour; archetypal information is stored in a typical mnemone; in biology, this mnemone is the metaphysical correlate of biological type/ super-species; it amounts to the *potential*, that is, blueprint or codified meaning behind physically expressed form; abstract or metaphysical precursor; the collective unconscious of a type e.g. human type;as thought is father to the deed or plan is prior to ordered action, so archetypes *precede*, hierarchically and temporarily, physical phenomena; pre-physical initial condition of matter.

ATP: Adenosine TriPhosphate, life's standard bearer of chemical/ heat energy; a cell's agent of energy transmission; a biological 'match' or 'battery'; an active cell may discharge many thousand units of *ATP* per second to drive its metabolic machinery; these are recharged by respiration; *ATP* also plays a critical informative role in the transmission of nervous and possibly other signals.

AV: Authorised Version e.g. *AVS* authorised version of science.

B

base: significant component of a *DNA* nucleotide: a letter in 'the

	book of life': there are 4 bases in the genetic alphabet - A (adenine), G (guanine), C (cytosine) and T (thymine): in the case of *RNA* base T is replaced by U (uracil).
base pair:	the conservative accuracy of genetic inheritance and the elegant construction of *DNA* are both dependent on a base-pairing rule viz. G pairs only with C and A with T (or U).
big bang:	*see* **transcendent projection**
biomimetics:	also known as biomimcry; fast-expanding field of scientific study and imitation of the biological production of codified substances and processes; research in order to better inform the processes of design (engineering) and technology (production) for human purposes.
black box:	process or system whose workings are unknown.

C

caduceus:	staff of Hermes/ Mercury the communicator, intercessor and informant deity; the messenger of metaphysic, carrier of thought, is a power mythologically trivialised; symbol, including double helix, used by Natural Dialectic to represent basic human infrastructure, that is, the archetypal form of man.
chaos:	a confusing notion with three main but disparate implications - emptiness, disorder and randomness; Greek word meaning chasm, emptiness or space; structureless 'profundity' that pre-existed cosmos; *prima materia, prakriti* or primordial energy structured by regulation of divinity, archetype or natural law; anti-principle of cosmos i.e. disorder; any case of actually or apparently random distribution or unpredictable behaviour; also apparently random but deterministic behaviour of systems (e.g. weather, electrical circuits or fluid dynamics) sensitive to initial conditions.
chloroplast:	organelle in plant cells containing photosynthetic apparatus.
chromatin:	a nuclear complex that, using histone proteins, helps package, reinforce and control the expression of genetic *DNA*.
chromosome:	a 'book' in the 'encyclopaedia' of life; the human genome contains 46 chromosomes.
cladistics:	method of classification using diagrams called cladograms; organisms are collected into groups on the basis of shared (homologous) features; homologies are tallied and numerical rather than speculative, evolutionary/ phylogenetic links drawn up between organisms; cladism is thus a powerful, neutral, objectively detached tool of analysis and for this reason the technique enjoys growing popularity among the world's taxonomists.
code:	the systematic arrangement of symbols to communicate a meaning; code always involves agreed elements of morphology (the form its symbols take), syntax (rules of arrangement) and semantics (meaning/ significance); without exception such prior agreement between sender (creator/ transmitter) and recipient involves intelligence.
codon:	a 'word' in the genetic language; stands for an amino acid or a full stop; since more than one codon may stand for a single

amino acid the genetic code is sometimes ill-perceived as 'degenerate'.

conscio-material dipole: illustrates the basic components of polar existence; informative and energetic components may be graphically modelled (as *fig.* 2.5) on vertical y- and horizontal x-axes respectively; the origin of such graph (zero information and zero energy) represents the cosmic sink - an abyss of space); sources of the couple extend from infinity towards this sink; Archetypal Potential scales from Psycho-Logical, Informative First Cause through grades of mind to non-conscious zero (matter); and potential matter (first cause physical) drops from original concentration of ultra-heat to, again, zero; thus cosmos is viewed as the gradual embodiment of Uncreated Source; it represents a scale of possibilities expressed as typical yet individual forms; by this token embodied soul is subject to individual incorporations (psychological and physical); these relatively dynamic forms constitute its psychological and, in the biological case, bodily circumstance; Natural Dialectic simply models such a hierarchical description of polar creation by the use of spectrum, concentric rings and, step-wise, ziggurats (see also cosmos and creation below).

convergence: the tendency of unrelated organisms to evolve similar characteristics; in the case of *divergence* adaptation/ speciation from an original feature occurs (e.g. beaks of finches); *convergence*, involving the unrelated, mosaic occurrence of similar features (such as the camera eye, viviparity and thousands of other instances), runs counter to Darwinian expectation; it means that such codified features must have evolved independently many times over; evolutionary explanations of this profound yet ubiquitous puzzle may thus involve speculations such as appeal to non-random 'deep bio-structure', 'principles of evolution', 'morphological laws' or 'inevitablility' granted by imaginary natural laws of codification/ innovation; for a design theorist the bio-codification and engineering of 'convergent' forms derives from either an original use of modular programming or, in the case of so-called micro-evolutionary variation, from in-built adaptive potential flexibly but appropriately activated by genetic switches and epigenetic markers.

cosmic fundamentals: cosmic psychological and physical qualities; basic states or tendencies; universal ingredients whose mixture is variously expressed in every object and event.

cosmological axis: human pivot; the point at which subjective and objective perception meet; eye-centre; third eye; thought centre; *ajna chakra.*

cosmological principle: idea that, on a sufficiently large scale, the distribution of matter in the physical universe looks just about the same from any vantage point; it therefore has neither centre nor, being infinite, edge - unless of course, its space is somehow spherical.

cosmos: physical universe, universal body; denotes orderly as opposed to chaotic process; involuntary pattern of nature; also equated, including metaphysical mind, with existence as a whole; seen,

	dialectically, as a projection through the template of metaphysical archetypes.
creation:	origination; physical or psychological arrangement; mind creates with purpose, matter without; creation means active production but also passive result; a creation will have been informed by force of mind and/or matter.

D

dialectical stack:	stack of opposites; columnar expression of polarity; there are two kinds of stack - primary or non-vectored and secondary, vectored; primary (essential) stacks set (*Sat*) Unity against (↓ *tam*/ *raj* ↑) duality (for elaboration see especially Chapter 2 and *figs*. 1.4, 2.2, 24.1 and 24.2); secondary (existential) stacks represent the various kinds of polarity from which the changeful web of existence is composed; each pair of polar 'anchor-points' implies a scale or dynamic range that runs between 'paired opposition' or 'complementary covalency'; stacks do not necessarily list synonyms or make equations; ***their perusal is intended to promote connections because consideration of connections tends to help unify/ collate/ organise one's working comprehension of any matter in hand.***
diploid:	having full genetic complement with one copy of chromosomes from each parent e.g. you have 46 chromosomes, 23 from mother and 23 from father.
DNA:	a complex chemical; a large bio-molecule made of smaller units, nucleotides, strung together in a row; a polymer in the form of a double-stranded helix; a medium superbly suited to the storage and replication of 'the book of life'; 'paper and ink' on which the genetic code is inscribed; an organism's 'hard drive'.

E

electromagnetism:	physics of the field that exerts an electromagnetic force on all charged particles and is in turn affected by such particles: light/ e-m radiation is an oscillatory disturbance (or wave) propagated through this field; light; light paradoxically involves a perfect, polar balance between contractive/ magnetic and radiant/ electric components.
elementary particles:	science has discovered and, for the most part, experimentally verified, over fifty elementary particles; these are divided, in simple terms, into bosons (force carrying particles) and fermions (separate particles); bosons include photons (which mediate the electromagnetic force), gluons (which mediate the strong nuclear force), W and Z particles (which mediate the weak nuclear force), possibly gravitons (which mediate the gravitational force) and also possibly a Higgs boson (which may mediate a proposed mass-giving field); fermions include two main groups - six quarks and leptons (six electron/ neutrino types); derived from quarks are strongly interactive composites called hadrons; hadrons include baryons

such as protons and neutrons and (perhaps a little confusingly) bosons such as short-lived mesons.

entropy: a measure of the amount of energy unavailable for work or degree of configurative disorder in a physical system (see second law of thermodynamics); inertial aspect of an energetic, material or conscious gradient; diffusion or concentration gradient outward from source to sink; drop towards 'most probable' outcome i.e. inertial slack; a measure of disintegration or randomness; expression of the (*tam* ↓) downward cosmic fundamental; a major property of matter, closely coupled with materialisation; in a closed system, which the universe may or may not be, this tends the eventual loss of all available energy, maximum disorder and the exhaustion of so-called 'heat death'.

enzyme: protein catalyst without whose type metabolism (and therefore biological life) could not happen.

epigeny: genetic super-coding; contextual punctuation; chemical modification of *DNA*; also extra-nuclear factors that may cross- reference with genetic expression.

equilibrium: three modes of equilibrium are (*sat*) balance of poise or pre-active potential; (*raj*) dynamic balance occurring in all regular cycles, wave-forms and cybernetic homeostasis that is basic to the stability of life-forms; and (*tam*) inertial equilibrium that results from diffusion of information or energy; it equates with exhausted inaction or 'flat', impotent rest; such post-active inertia represents the most probable distribution of energy/ matter with the least energy available for work viz. the most random arrangement permitted by the constraints of a system; expressed in psychological terms as ignorance, unconsciousness or sleep; see also equilibration, *karma* and *fig*. 3.3 'Pivoted Existence'.

Essence: (*Sat*) Supreme or Infinite Being; Substance (perhaps Spinoza's Substance) 'prior to' or 'above' existence; Pure Consciousness/ Life; Peace that transcends all psychological and physical action; the root of an essentially undivided universe; Uncreated One within which and whence all differences have their being; Apex of Mount Universe; goal of saints/ 'philosopher kings'; the 'point' at which All-Is-One.

eukaryote non-prokaryote; any organism except bacteria and blue-green algae.

evolution: there are today *four* main usages of this word; each 'loading' derives from the original Latin, 'evolvere', meaning to unroll, disentangle or disclose; the *first two*, physical and biological, are conceived as natural/ mindless processes; the *second, mindful pair* is of psychological/ teleological import; specious ambiguity may conflate or switch between the fundamentally separate pairs of meaning. **Firstly,** in the scientific context of physics and chemistry, the word is used to describe change occurring to physical systems; the laws of nature can't, it seems, evolve through time but stars, fires, rocks or gases can. **Secondly,** though also subject to the 'rules' of entropy, biological evolution is a theory of *random progression* from simple to complex form; it thereby implies increasing, codified complexity; while retaining the 'hard loading' of physical science it also, ambiguously, claims that codes, programs, mechanisms and coherent, purposive systems - normally the province of mental concept - self-organise by, essentially, chance; such confusion, the basis of naturalism, is compounded by failure to distinguish between, on the one hand, ubiquitously observed variation (called micro-evolution) and, on the other, Darwinian 'transformation' between different sets of body plan, physiological routines and associated types of organism - such 'black-box macro-evolution' as is never indisputably observed; to evoke a naturalistic ambience it is fashionable to use 'evolved' interchangeably with or to replace the words 'was created', 'was planned' or 'designed'; finally, it is noted that the coded, choreographed development of a zygote, packed with anticipatory information, through precise algorithms to adult form is the absolute antithesis of blind Darwinian evolution. *Thirdly*, man certainly evolves ideas; intellect can evolve 'purposive complexity'; we invent all kinds of codes, schemes and machines; we devise increasingly complex theories and technologies; and we evolve an understanding of natural principles; this, which all parties accept, is an informative, psychological sense of 'evolution'. The *fourth* sense of evolution, at least as near to the original Latin as the other three, is the spiritual usage; immaterial spiritual evolution, unacceptable to materialists and unknown to physical science, is at the very heart of holism; in this voluntary sense of evolution practitioners cast off material attachment, evolve and merge into the *Logos*; evolution (or, perhaps better, centripetal involution) of the soul is their great business; their aspiration is to unite with The Heart of Nature.

evolution pre-Darwinian: minority/ anti-mainstream pre-Socratic snippets and sense-based Epicureanism lionized by interpretations of post-18th century materialists; virtually undetectable eccentricity in Chinese, Indian and Islamic literature; natural selection treated by creationists al-Jahiz and Edward Blyth; Buffon, a non-evolutionist, addressed 'evolutionary problems'; Lamarck (evolution by inheritance of acquired characteristics); hints in poem by Erasmus Darwin.

evolution Darwinian: mechanism - natural selection; major tenets - common descent (inheritance), homology and 'tree of life'.

evolution: neo-Darwinian/ synthetic: as Darwinian, except synthetic theory adds random mutation as the mechanism for innovation; also adds a mathematical treatment of population genetics and various elements (e.g. geno-centric perspective) derived from molecular biology.

evolution: post-synthetic phase: natural selection and random mutationare acknowledged as mechanisms insufficient to source bio-information; post-Darwinian evolution invokes mechanisms from hypotheses such as *NGE* (natural genetic engineering) and 'evo-devo'; holistic possibilities also address the origin of complex, specified and functional bio-information.

existence: which 'stands out' from background 'nothingness'; the apparently divided universe; seemingly disparate, finite things; all motion/ change/ relativity; all psychological and physical events.

exon: specifies the amino acid sequence for a protein; m-*RNA* after protein editors have removed introns.

F

field: any extent wherein action either physical or metaphysical but of a certain kind occurs e.g. field of battle, influence of mind or magnetism; the scientific definition is limited to a collection of numbers varying from point to point - such as a scalar field of contours on a map - or numbers with direction - such as a vector field showing speeds and directions of wind.

first causes: check *figs.* 7.2, 14.1 or 14.3. First cause is first motion in a previously undisturbed, pre-conditional field. **First Cause Psychological** is Archetype, Potential Informant or (see Chapter 5: Top Teleology) *Logos*; attributes of this Primary Source and Sustenance of Creation include omnipresence, omnipotence and omniscience; **first cause physical** is also called potential matter or archetypal memory; as the secondary source of creation it precedes physical phenomena; as such it is, transcending physical appearances, metaphysical; this 'physical nothingness' is therefore, paradoxically, the source of everything composing astronomical cosmos; it consists of their being or essence as opposed to their becoming; its void, with respect to the presence of finite phenomena, appears infinite; attributes of immanent archetype, the primary informant of our non-conscious, energetic universe, include omnipresence and omnipotence.

G

gamete: sex cell with half of full genetic complement i.e. a single set of chromosomes.

gene: generally means a basic unit of material inheritance; section of chromosome coding for a protein; digital file; a reading frame that includes exons and introns; the old one gene-one protein hypothesis is incorrect; in fact, by gene splicing, a particular piece of *DNA* may be used to create multiple proteins.

genome: total genetic information found in a cell: think of the genome as an instruction manual for the construction and physical operation of a given organism

genotype: the genetic constitution of an organism, often referring to a specific pair of alleles; the prior information, potential, plan or cause of an effect called phenotype.

gravity: in physics an attractive mass-to-mass force or warping of space-time; in Natural Dialectic the term is redefined more broadly - the agency of its (*tam* ↓) downward vector includes all psychological and physical factors of materialisation; such 'gravitational' factors and their properties are listed in the left-hand column of Secondary, Existential Dialectic; they include pain, pressure, confinement, strong nuclear force, mass, electromagnetic binding, inertia, entropy, 'standard' gravity and so on; gravity might be summarised as 'negative power' or 'the principle of death'.

H

haploid: having half the full genetic complement, as in the case of sex cell.

heterozygous: having different allelic forms of a particular gene.

holism: opposite of reductionism; the view that a whole is greater than the sum of its parts; the extra metaphysical (immaterial) ingredient is identified by Natural Dialectic as information; information implies the purposeful design, development and arrangement of contingent parts in a working system; may operate according to a Logical Norm.

hologram: a 3-d photograph made with the help of lasers. Unlike a normal photographic image each part of it contains the image held by the whole.

homeostasis: vibratory or periodic control of a system to obtain balance round a pre-set norm; the mechanism of its information loop involves sensor, processor and executor; the operative cycle works by negative feedback; psychological (nervous) and biological cybernetics; the informed basis of biological stability.

homeotic gene: gene involved in developmental sequence and pattern; high-level co-determinant of the formation of body parts.

homozygous: having the same allelic forms of a particular gene.

I

illusion: is the cut between illusion and delusion an illusion? illusions, apparently outside the mind, appear real; a delusion, in it, we think real; neither, mind allows, is real or true.

information: the immaterial, subjective element; information is action's precedent; informative potential is both a psychological and (by way of archetype) physical entrainer; information is the inhabitant of its own centre, mind, whose substrate is consciousness; active information knows, feels, purposes and codifies; it recognises meaning; on the other hand passive information reflects active; it is stored as subconscious memory; or is fixed in the expressions of non-conscious matter according, universally, to the archetypal behaviours of natural bodies or, locally, to particular constructions by life-forms.

informative entropy: loss of information due to degradation of its carrying medium; such a medium may be metaphysical (mind) or passive and physical (for example, computer files or genetic code); and its entropy may be metaphysical (loss of memory, focus or consciousness) or physical (for example, genetic mutation); the informative correlate of such degeneration is diminished organisational capacity, meaning or thrust of original purpose.

informative negentropy: gain of informative clarity; increasingly focused, purposive specificity; associated with knowledge, wisdom, grasp of principle and pristine construction.

intron: genetic control panel; n-p-c (non-protein-coding) segment(s) spliced from an m-*RNA* transcript prior to translation; introns

	include regulatory elements (to variably promote or inhibit gene expression) and addressing factors of the genetic operating system; gene-attached information lending specific flexibility to protein manufacture.
inversion:	turning upside-down or inside-out; reversing an order, position or relationship; in a hierarchical sense inversion is allied with the reflective asymmetry of opposite poles; information outwardly expressed; pole-to-pole reversal integral to dialectical structure; various kinds of inversion (cosmic and micro-cosmic (or biological)) are discussed.

K	
karma:	action; law of cause and effect, that is, balance between action and reaction; equilibration such as underlies all mathematical equation; a deed with implications of the reactions or 'payback' it provokes; fruit or result of previous thoughts, words and deeds; applies as rigidly to metaphysical (psychological) as, in Newton's Third Law of Motion and mathematical *equations*, physical events.

L	
lepton:	*see* **elementary particle**
levity:	agency of the (*raj* ↑) upward vector; dialectical converse of gravity; psychological and physical 'levitatory' forces lift or stimulate; they are listed in the right-hand column of Secondary, Existential Dialectic and include light, heat, excitement, dematerialisation, release, negentropy, focus of interest, affection and so on; physically, levity includes anti-gravity or the intrinsic property of matter's absence, space; generally summarised as 'positive power' or 'a buoyant principle of liveliness'.
logic:	analysis of a chain of reasoning; principles used in circuitry design and computer programming; 'normative reason' relates to the basic axiom(s) of a given standard e.g. *bottom-up* materialism or *top-down* holism; three main logical thrusts are: (1) inductive (premises/ observations supply evidence for a probable/ plausible conclusion) as in the case of experimental science working *bottom-up* from specific instances to general principle: (2) abductive (best inference concerning an historical event): and (3) deductive (conclusion in specific cases reached *top-down* from general principle): two pillars of logic are holism and materialism; holism employs mainly deductive/ abductive operations and a Logical Norm; materialism tends to inductive/ abductive operations whose axis is non-conscious force and chance.
Logos:	First Cause; Prime Mover; Causal Motion that sustains creation's conscio-material gradient.

M	
macrocosm:	the physical universe of astronomy and cosmology; dialectically, the whole of existence (i.e. both universal mind and universal body) as opposed to individual, microcosmic objects and events - including the human body.

macro-evolution: large-scale, non-trivial evolution; process of common or phylogenetic descent alleged to occur between biological orders, classes, phyla and domains; includes the origin of body plans, coordinated systems, organs, tissues and cell types; unexplained by mutation, saltation, orthogenesis or any known biological mechanism; sometimes called 'general theory of evolution' (*GTE*); macro-evolution, an extrapolation from Darwinian micro-evolution vital to sustain a 'progressive' materialistic mind-set, is conjecture.

matter-in-practice: bulk, bonded matter including all molecular-based substances; gross matter; external appearance.

matter-in-principle: quantum phase; particles and forces; subtle matter; internal cosmic drivers.

meiosis: shuffling the information pack: variation-on-theme; mechanism for the production of haploid gametes; genetic postal system for sexual reproduction.

metabolism: body chemistry.

metaphysic: = non-physical/ immaterial/ psychological/ unnaturalistic; physically expressed as specific/ intended arrangement/ behaviour of materials; physical behaviour reflects metaphysical blueprint; involves element of information; also involves symbol/ code/ abstraction/ logic/ reason/ mathematics; also message/ meaning/ goal/ teleology; also consciousness/ mind/ life/ experience/ feeling; and also morality/ force psychological/ emotion; involves innovation/ creativity/ art/ invention/ aesthetics.

microcosm: an entity that reflects the universe by containing all its basic constituents. Used especially of the human state where it may refer to both mind and body or, in a purely physical context, body alone.

micro-evolution: misnomer; non-progressive, small-scale variation within a species or, more broadly, between strains, races, species and genera; *variation/ adaptation within type*; trivial Darwinian changes that may occur by natural selection/ ecological factors acting on genetic recombination or mutation; sometimes called 'special theory of evolution' (*STE*), micro-evolution/ variation is a fact.

mitochondrion: organelle in eukaryotic cells containing the apparatus for aerobic respiration.

mitosis: conservative copying and delivery of genomes in cell division; genetic reprinting; genetic postal system for asexual reproduction.

mnemone: a division of memory whether individual or universal: an individual's two divisions are *personal mnemone* (likened to a working cache or data store) and *typical mnemone* (likened to a *ROM* or an operating system); typical mnemone is, in effect, a program consisting of three subroutines - *signal translation, instinct* and *morphogene* (for more information see *SAS* Chapters 15 - 17); it is also a synonym for natural, archetypal memory in universal mind; in short, it is a body's ***metaphysical DNA***. Further than the character of each bio-

mobile genetic element: transposon, retrotransposon, insertion sequence, other non-protein-coding *DNA*, n-p-c *RNA* fragments and various protein regulators that together expedite the operating system of a genome.

morphogene: one of three sub-routines of typical mnemone or archetypal one of three sub-routines of typical mnemone or archetypal memory relating to physical construction; morphological attractor; the component of subconscious mind associated with electrochemical function and thereby body; just as you might not guess from the picture on your TV screen or object from a 3-d printer the nature of the electromagnetic messaging that creates it so you might not guess a body's shape from its *DNA* or the messaging agent that links archetypal mind with body; morphogene is the dominant, perhaps exclusive, aspect of mind in unconscious organisms such as plants or fungi.

morphogenesis: the development of biological structure; more generally, the production of physical form.

mosaic: the presence of permutations of codified sub-routines or similarities of form and/or function scattered in organisms unrelated by lineage.

mutation: accidental change to genetic code.

mysticism: quite different from objective, it is the subjective science; not philosophy, religion or opinion but practice to achieve communion with natural, inner, immaterial truth; esoteric as opposed to exoteric, materialistic discipline; 'science of the soul'; as gyms and physical action are to athletes so meditative exercise and psychological stillness are to mystics; involves psychological techniques to achieve a clear, rational goal - purity of consciousness and thereby understanding of the fundamental nature of the informative principle, mind; since life is lived in mind a mystic seeks consummate knowledge of life's source and sanctum, that is, communion with its deathless heart; adepts were, are and will be 'Olympian' meditative concentrators.

N

nano-biology: biology of structures/ physiologies involving a few atoms or molecules; 'extremely small biology'.

nanotechnology: technology at atomic and sub-atomic level as is, basically, life's.

natural law: the automatic, reflex and mathematically describable behaviour of a physical entity; likewise the repetitive nature of its interactions with other entities.

naturalistic methodology: also known as 'methodological naturalism', this strategy is, strictly, not concerned with claims of what exists or might exist, simply with experimental methods of discovering physically measurable behaviours; thus only materialistic answers to any question (e.g. how biological

	forms arose or the nature of mind) are deemed 'scientific' or 'scientifically respectable'.
negentropy:	opposite of entropy; lowering of entropy; expression of the (*raj*) upward-pointing cosmic fundamental closely coupled with stimulus, dissolution and dematerialisation; a measure of input, cooperation or synthesis; motive/ fluidising aspect of an energetic, material or conscious gradient; gain of energy, configurative order, information or consciousness in a system; when used in terms of information negentropy involves gain in order or understanding of principle from which different actualities derive; a measure of the amount of concentrated/ conceptual information, specific, intentional complexity or conscious arrangement in a system; a natural and essential property of mind.
Nirvana:	state of enlightenment; 'non-condition'; *nirvana* is devoid of existential motion; extinction of existence leaving Essence Alone; pure soul; psychological super-state; Buddhists call such transcendence non-self or the Formless Self.
non-existence:	where creation = formful existence, non-existence is formless; the polar opposite of physical space and time is Transcendent Potential; such pre- or super-existential formlessness is non-existent; Absolute Non-Existence is Essential; however relative non-existences of two kinds also occur; the first kind is metaphysical/ subjective and therefore psychological; it involves the absence of a specific psychological form or event; unconscious oblivion is one such non-existence; the second kind involves the local absence of a possible physical event (an object is a 'slow event'); impossibilities are non-existences but imaginations of non-existence (including symbolic abstractions, hypothetical entities, physical absences, absolute emptiness and the number zero) exist; furthermore, the nothingness of space and time, the zero-point of calculus and zero's empty set together constitute the basis of physical science and mathematics.
non-protein-coding *DNA*:	occupies probably 95% of eukaryote and 80% of bacterial genomes; associated with the genetic operating system; may include some genuinely redundant misprints or duplications but now thought for the most part critical to the flexibility, efficiency and even possibility of gene expression; once thought of as useless, degraded information and ignorantly called 'junk *DNA*'.
non-protein-coding *RNA*:	n-p-c *RNA* is also called nc-*RNA* (non-coding), nm-*RNA* (non-messenger) or f-*RNA* (functional); functional *RNA* molecule not translated into protein; many 1000's of different specimens include classes of t-*RNA* (transfer *RNA*), r-*RNA* (ribosomal *RNA*) and, commonly involved in the regulation of gene expression and other intra-cellular tasks, micro-*RNA*, double-stranded si-*RNA*, pi-*RNA* and so on; also, for inter-cellular communication, ex-*RNA*.
nucleic acid:	see *DNA* and *RNA*

nucleosome: a 'reel' composed of histone proteins around which chromosomal *DNA* is precisely wrapped; repeated nucleosomes allow the *DNA* to form a bead-like structure that can coil and super-coil; *DNA*, nucleosomes and other factors compose chromatin.

nucleotide: basic, triplex unit of nucleic acid polymer; monomer composed of phosphate and sugar (the 'paper' part) and base (the 'ink letter'); letters' of the genetic alphabet are (G) guanine, (C) cytosine, (A) adenine and (T) thymine. In *RNA* thymine is replaced by (U) uracil.

O

object: a slow, although energetic, event; apparent fixation.

Om: universal sound, fundamental reverberation, basic truth; sometimes spelt *Aum*, a Sanskrit word whose Semitic transliterations are Am'n, Amin and Amen; see also First Cause, *Logos*, *Kalam*, *Shabda* etc.

order: regular, regulated or systematic arrangement; organisation according to the direction of physical law; passive information by which things are arranged naturally (with predictable but non-purposive complexity) or purposely (with innovative or specified complexity); mind, generating specified complexity in the order of its technologies and codes, actively informs; the orders of mind are meaningful, the orders of matter lack intent.

organelle: cellular sub-station; discrete part of a cell; sub-cellular compartment having specific role such as informative (nucleus), energetic (mitochondrion, chloroplast), constructional (ribosome, Golgi body) or other.

P

***PAM, PAND, PCM* and *PCND*:** philosophical gambits; see Primary Axioms and Corollaries.

phenotype: the effect of causal potential; result of the development of prior, informative 'egg'; outward expression of inner plan; sensible appearance of an organism as opposed to its genotypic scheme: the whole set of outward appearances of a cell, tissue, organ and organism are sometimes called a phenome (*cf.* genotype/ genome).

photosynthesis: process by which inorganic carbon is introduced to the biological zone and energetic sunlight fixed as a crystalline molecule of storage, a sugar called glucose.

phylogeny: evolutionary history; relationships based on common or evolutionary descent.

potential: poise; latent possibility; potent non-action that precedes any particular action or creation; in science potential energy is defined as the energy particles in a system (or field) possess by virtue of position/ arrangement; gravitational, electrical, electro-chemical, thermo-dynamical and other kinds of potential are recognised; in dialectical terms mind precedes matter, information precedes the pattern of material behaviour; information is energy's pre-requisite potential; in this case *informative potential* involves two conditions; firstly, a pre-existential/ essential state of pure potential; secondly, a pre-material, metaphysical fact of potential matter, archetype or laws of nature; if potential's pre-active equilibrium is related to the voltage of a full battery then aspects of psychological 'voltage', whose currents drive intentional behaviour, are purpose, will and plan.

potential matter: see archetype.

Primary Axiom of Materialism (*PAM*): all objects and events, including an origin of the universe and the nature of mind, are material alone; cosmos issued out of nothing; life's an inconsequent coincidence, a fluky flicker in a lifeless, dark eternity.

Primary Axiom of Natural Dialectic (*PAND*): there exists a natural, universal immaterial element - information; immaterial informs material behaviour; a conscio-material dipole that issues from First Cause informs and substantiates both mental (metaphysical) and physical creations; there is eternal brilliance whose shadow-show is called creation.

Primary Corollary of Materialism (*PCM*): the neo-Darwinian theory of evolution, that is, life forms are the product, by common descent, of a random generator (mutation) acted on by a filter called natural selection; such evolution is an absolutely mindless, purposeless process; the *PCM* is a fundamental *mantra* of materialism.

Primary Corollary of Natural Dialectic (*PCND*): the origin of irreducible, biological complexity is not an accumulation of 'lucky' accidents constrained by natural law and death; forms of life are conceptual; they are, like any creation of mind, the product of purpose.

prokaryote: non-eukaryote; bacterial type with little or no compartmentalisation of cell functionaries.

promissory materialism: belief sustained by faith that scientific discoveries will in the future justify/ vindicate exclusive materialism and, as a consequence, atheism; may involve a call to progress towards the technological provision of its 'promised land'.

protein: factor made from a specific sequence of amino acids to perform a specific task; 'informative' protein includes some hormones; skin, hair, bone, muscle and other tissues are made of 'structural' protein; 'functional protein' called enzymes mediates all stages in cell metabolism, that is, it catalyses all biochemistry.

***PSI* (psychosomatic interface):** psychosomatic border; the level of mind-matter interaction; bridge between metaphysical and physical dimensions; potential matter; 'gap of Leibniz'; 'fit' of mind to matter; point of linkage between subconscious mind and non-conscious matter; gearing between instinct/ archetype and the behaviour of material objects and energies; as in the case of physical law, psychosomatic influence is both general in potential and local/ specific in engagement.

psychological entropy: a measure of loss of concentration, focus of attention or consciousness; loss of 'mental energy' or aptitude; the drop from waking to sleep; loss of knowledge, information or sensitivity; the gradient from intelligence through stupidity to oblivion; an expression of the (*tam*) downward cosmic fundamental in mind; a tendency predominant in lower, egotistical or selfish mind; increasing level of ignorance, anguish or immorality; loss of integrity, psychological disharmony or disintegration; see also *information entropy*.

psychological negentropy: a measure of gain in order; an increase in concentration, focus of attention or consciousness; gain in sense of purpose, 'mental energy' or aptitude; the rise from sleep to waking, 'dark to light' or unhappiness to happiness; gain in knowledge, information or sensitivity; the gradient of learning and spiritual evolution; an expression of the (*raj*) upward cosmic fundamental in mind; a tendency predominant in higher mind; increasing level of contentment, understanding and the natural morality of happiness; the ascent towards psychological radiance, harmony and integration. The converse of psychological negentropy involves *entropy of information*.

psychosomasis: you are a *psychosoma* (mind-body entity) and it is the interface between these two elements that interests us; *psychosomasis* is operation across the psychosomatic border; mind/ body interaction; the one-way, morphogenic imposition of archetypal pattern on *physicalia*; and the two-way exchange of information in sentient organisms through the agency/ medium of subconscious patterns.

Q

quantum: minimum discrete amount of some physical property such as energy, space or time that a system can possess; quantum theory states that energy exists in tiny, discontinuous packets each of which is called a quantum; an elementary discontinuity; an elementary particle e.g. photon or electron.

quantum level: matter-in-principle; 'internal', 'causal' or 'subtle' matter; the vibrant or energetic phase of physical organisation; zone of sub-atomic particles and forces; step (on cosmic ziggurat) between potential and bulk matter whose aspect is sometimes extended to include atomic and molecular interactions; small-scale substance underlying large-scale, sensible appearances.

R

raj: (↑) upward, levitatory or stimulatory cosmic vector.

reductionism: opposite of holism; the materialistic view that an article can always be analysed, split up or 'reduced' to more fundamental parts; these parts can then be added back to reconstruct the whole; a whole is no more than the sum of its parts.

religion: etymology debated between Latin *religare* (bind) and *relegere* (review); *religio* means dutiful and meticulous observance; currently religion means world-view, mind-set or basic faith; whether of materialistic or holistic belief, it involves the non-negotiable substance of an individual or community's truth - notably as regards origins; antagonism between holistic practice and the naturalistic methodology of science is,

	because the couple deal with separate but complementary physical and metaphysical dimensions, flawed; a materialist/ atheist 'binds meticulously' to an evolutionary mind-set, a holist to pantheism or a Living Creator; in the case that self-deception is crucial to successfully deceiving others which, holism or materialism, is the religion that is ultimately true?
resonance	the tendency of a body or system to oscillate with a larger amplitude when subjected to disturbance by the same frequencies as its own natural ones; thus a resonator is a device that naturally oscillates at such (resonant) frequencies with greater amplitude than at others; resonance phenomena occur with all kinds of vibration, oscillation or wave; their sorts include mechanical, harmonic (acoustic), electrical (as with antennae), atomic and molecular.
respiration:	the controlled release of energy from food.
ribosome:	site of polypeptide (protein) synthesis.
RNA:	a single-stranded nucleic acid polymer employed in three different forms during the process of protein synthesis; in computer terms might be likened to a portable memory stick as opposed to *DNA*'s hard drive.
m-RNA	is used to transcribe a base sequence from *DNA*. It 'photocopies' a gene and carries this information to a ribosome.
micro-RNA	short mi-*RNA* molecules are important regulators of genetic expression.
r-RNA	is part of the make-up of the protein-manufacturing station called a ribosome.
t-RNA	critically translates genetic 'words' (see 'codon') into amino acids: 64 such operators form the link between code and the actuality of a functional protein.

S

sanskara:	character trait; groove, habit, obsession or repetitious mode of thought proportional in depth to the intensity of desire, force of impact or impression that created or sustains it.
sat:	'top' or essential cosmic fundamental; 'vector' of balance, neutrality.
science:	Latin *scire* (know); knowledge; commonly understood as the practical and mathematical study of material phenomena whose purpose is to produce useful models of the physical world's reality.
scientism:	a philosophical face of official, *de facto* commitment to materialism; today's majority consensus of what the creed of science is; an -ism born of *PAM*; a faith that all processes must be ultimately explicable in terms of physical processes alone; like communism, a one-party state of mind; a doctrine that physical science with its scientific method is ultimately the sole authority and arbiter of truth; a set of concepts designed to produce exclusively material explanations for every aspect of existence, that is, to colonise each academic discipline and build its intellectual empire everywhere; 'scientific fundamentalism' closely allied, when expressed in social and political terms,

with 'secular fundamentalism', sociological interpretation of behaviour and the fostering of a humanistic curriculum.

secular fundamentalism: *PAM* as applied to the worlds of nature and of human society.

secularism: concern with worldly business; lack of involvement in religion or faith; secularism is generally identified, as defined by the dictionary, with materialism; for a secularist the ultimate arbiter of truth is human reason - ideas are open to negotiation so that even morality is relative; however many liberal agnostics, atheists and humanists argue that their metaphysical, philosophical system also embraces so-called 'universal' moral values and, as opposed to zealotry or the logic of evolutionary faith, a liberal politic of 'philosophical live-and-let-live'.

stereo-computation: stereochemistry involves study of the relative spatial arrangement of atoms in molecules; in biology a 1-D line of informative code (whose 3-D constituents bear no figurative relationship with their informed product) give rise to relative 3-D spatial arrangements at all levels from molecular to systemic and whole-body. Such targeted generation may be termed bio-logical stereocomputation.

sub-state: *opp.* super-state; impotence, discharge, exhaustion, final stage in the expression of potential; fixity; non-conscious base-state; state 'below/ subtendence; extreme negativity/ (*tam*) condition.

super-state: potential; source of possibility; causal metaphysic/ archetype; state 'before' or 'above' subsequent expression; immanence; transcendence; precondition; (*sat*) priority.

symmetry: an aspect of the (*sat*) characteristic of balance; aesthetically pleasing balance and proportion; geometrical balance or interactive process such that some feature of an action remains invariant, that is, conserved; the symmetry of an entity (such as a sphere, empty space or natural law) or feature (such as energy) that remains the same at all times everywhere from any local point of observation or through every transformation is called 'higher' or 'continuous'; if a feature is conserved only when an object or process is moved, turned or viewed at certain angles or under specific conditions its symmetry is called 'lower' or 'discrete'; the symmetrical properties of a system may be precisely related to corresponding conservation laws and *vice versa*; internal symmetries found in quantum physics (such as gauge transformations) are independent of space-time coordinates; scale symmetry occurs when a reduced or expanded object keeps its shape but not its size (as with Mandelbrot fractals); dialectical symmetry also involves *informative potential*; its metaphysical archetypes inform principles, laws or determinant fields that exist prior to action and, from their possibilities, govern actual outcome; such 'configuration of the world' is absolute and, beyond entropy, stable; it is negentropically immune from decay; by contrast, the 'free' symmetry of potential energy is inherently unstable and (like a pencil balanced on its tip) liable to spontaneously 'topple' or 'break' into the least energetic of a range of circumstantial possibilities; such spontaneous symmetry-breaking, the basis of diversity, represents an expression of 'deep symmetry' or archetype under local conditions and is therefore called by physicists 'contingent'.

T

tam: ($\downarrow$) downward, gravitational or inertialising cosmic vector.

teleology: the doctrine that there is evidence of purpose in nature; doctrine of non-randomness in natural architecture; doctrine of reason ('for the sake of', 'in order to', 'so that' etc.) and intent behind biological and universal design.

third eye: place where you think; point of metaphysical focus between and behind the eyebrows, that is, just above the physical eyes; HQ/ seat of mind beyond the sensory world; cosmological eye-centre; gate through which meditative concentration can pass; single way that leads within.

transcendent projections: psychological: see Index: *Logos*, archetype and *figs*. 7.1, 11.2 and/ or 11.3; also *SAS* Chapter 5: Top Teleology and *figs*. 9.1 and 11.1.

physical: see Chapter 14; Glossary: archetype; Index: transcendence, archetype, cosmo-logical language; *figs*. as above and 14.3; such projection involves an orderly, energetic expression from either metaphysical or physical nothingness, that is, unseen potential; an instantaneous 'miracle' that issues from 'within' non-conscious physicality; transcendently emergent, finely tuned expansion from 'inner' metaphysic into 'outer' material/ natural law; 0-dimensional singularity from whose prior pointlessness all points perhaps began; cosmic seed whence, *ex nihilo*, the world developed; projection whose appearance, once physical, is visible and perhaps described but certainly not explained by big bang theory; transcendent projection of archetype is possibly, to the constrained sensory and intellectual states of human mind, ultimately incomprehensible; its invisible dynamic, the practice of materialisation, may remain a fact beyond material understanding. for references involving more detail about psychological, physical and biological projections see *SAS*: Glossary.

biological: if matter is developed memory (Chapter 14: Space) then see Chapters 13: Typical Mnemone and 15: Conceptual Biology; see also Glossary: mnemone and archetype; Index: mnemone, archetype; and *figs*. 7.1, 11.2, 11.3, 13.7 and 15.1.

transposon: 'jumping gene'; ubiquitous genetic element found in all prokaryotes and eukaryotes so far investigated; *DNA* segment that can, by enzyme, be cut from a one site (the donor) and joined to another (the target); a retrotransposon is moved through the mediation of *RNA* and reverse transcription back from *RNA* to *DNA*; transposons and retrotransposons play a key, functional role in gene expression and regulation; a kind of retrotransposon, *SINEs* and

*LINE*s are thought to compose 35+% of the human genome; such elements may be flanked by terminal repeats that allow specific, operational variation in different types of cell; from an evolutionary view they comprise functionless viral imports; from a *top-down* view it is predicted they form a dynamic, intrinsic element of the genome involved in gene regulation, genetic shuffling as (epigenetic) response to buffer circumstantial exigency and, as important, structural agents able to reshape a chromosome to meet specific genetic demands.

transcription factor: a protein that, binding to a specific *DNA* sequence, regulates genetic transcription.

truth: what's basic to or binds a group; substance; a universal truth substantiates, as source and sustenance, all things.

U

unification: progressive unification of forces is the grail of physics: Clerk Maxwell unified electricity and magnetism; electroweak or *GSW* theory brought in the weak nuclear force; now the goal is to include the strong nuclear force (*GUT*), gravity in a super-force and show that, in essence, particles and forces are interchangeable (super-symmetry and *TOE*); Natural Dialectic, also working with the maxim 'All is One', includes what sums to a hierarchical *TOP* or Theory of Potential (see especially Chapters 3, 4, 8, 10 and 11, also *figs.* 7.1-2); potential is the absolute from which variant orders of relativity derive; the subjective potential for mind is consciousness and the objective potential for matter is archetypal memory; such archetypal element unites psychology with the physics of natural science; it is the informative precondition of physical and biological form.

universal mind: cosmic grade; also called the 'mind of nature' or 'natural mind'; as a biological body is a specific though complex arrangement of universal chemicals so individual mind partakes of a particular, equally minuscule fraction of the metaphysical components of universal mind; *see* also archetype.

V

vector: existential dynamic; direction of travel with respect to model or secondary stack used in Natural Dialectic; fundamental vectors (↑ and ↓) denote relative gain or loss of information or energy; and, similarly, motion towards and from the axis, peak or source of cosmic model; general, metaphorical rather than specific use of the word; not, therefore, the same as that defined by physics or biology.

virtuality: exotic component of quantum physics; para-physical feature of the quantum vacuum; immaterial substrate of material phenomena; inner (where solidity's the outer) edge of physical reality; ephemeral 'virtual particles' rise and sink back into a 'void' thought to teem with their 'fluctuations'; virtuality is identified as the agent of such important actualities as the strong nuclear force (resulting from interaction between virtual mesons and gluons), vacuum polarisation, the Coulomb force (between electric charges and mediated by the exchange flight of virtual photons) and so on; not used in the computer sense of a continuum between real and imaginary circumstance; see also *ZPE*.

Z

zero: zero (the number) is a metaphysical entity, one critical to mathematics; zero (the fact) means, for Natural Dialectic, nothing in two senses; in the *negative sense* it means an absence of perception (psychological oblivion) or absolutely nothing physical (as naturalistically prescribed to precede, say, a big bang or as the nature of a theoretically perfect vacuum); negative sense may also be construed as (*tam*) an extreme sub-state, sink or emptiness; for materialism 'absolute nothingness' may involve natural law and its mathematical description; what, one may enquire, is the source of such 'eternal metaphysic', what is the nature zero-physical?; on the other hand, in a *positive sense* zero refers to source, pre-existent potential or (*sat*) higher cause-in-principle; for example, information (which is zero-physical) transcends/ precedes a course of action; information that passively governs the operation of cosmos derives from immaterial archetype.

ZPE: zero-point energy; quantum vacuum; vacuum energy of all fields in space; residual energy of all oscillators at 0°K; concept first developed by Albert Einstein and Otto Stern; intrinsic energy of vacuum; the ground-state minimum that any quantum mechanical system, in particular the vacuum, can have; remainder, according to the uncertainty principle, when all particles and thermal radiation have been extracted from a volume of space; residual non-thermal radiation; irreducible 'background noise'; 'quantum foam'; the potent, microscopic side of quantum vacuum (as opposed to impotent, macroscopic vacuum left by the apparent lack of anything); subliminal 'rumblings' of immaterial weak, strong and electromagnetic fields (called *ZPF*s); seething, jostling ferment of subliminal waves and particles in emptiness; a flux of unobservable 'virtual' matter and anti-matter that may or may not appear as the basis of observable forces such as electromagnetism, charge and perhaps inertial mass and gravity; a subtle facet of levity; the anti-gravity of dark energy (or the cosmological constant) has been postulated as a component of *ZPE*; suggested 'mother-field' support for electron orbits, atomic structure and thus the phenomenal universe.

zygote: fertilised egg.

Index

A

abiogenesis
  biochemical puddle 142
  chemical evolution 67, 142, 144
  life-by-matter 67
  spontaneous generation 142
absolute physical/ apparent absolute
  0K (zero degrees Kelvin) 50
  potential matter .... *see* potential/ archetype
  present absence/ basic physic/ nothingness of space & time ...... *see* space/ time/ space-time/ nothingness/ void
  source of physical form *see* archetype/ potential matter
  subtendence *see* subtendence/ negative power
  velocity of light (c) 51
Absolute Psychological
  (N)One/ The Infinite 40
  Cosmic Source of Relativity 24, 50
  Essence as *opp*./ beyond relative existence 55, 62
  Morality 152, 153
  Subjectivity/ Objectivity 77
  Top Viewpoint/ Value/ Criterion 40, 59, 148, 149, 152
  Truth/ Reality 147, 148, 149
act of creativity.... *opp*. act of perception; mind > matter/ materialisation of idea *see* also innovation/ creator
active complexity .... *see* purposive complexity
active information .... *see* information
adaptive potential 140
Adleman Leonard 135
Ain Sof .... 53, *see* also Essence/ (N)One
alchemy
  chemical evolution 141, 144
  information from matter. *see* creation stories/ innovation/ matter/ miracle/ Law of Non-Conscious/ Naturalistic Non-Innovation
Algazel (Al-Ghazzali) 54
Alpha
  Moment .... 77, 91, 153, *see* Consciousness
  Source 42
Alpha Moment
  Source 125
alphabet*see* also code/ language/ cosmic language/ genetic grammar/ quantum aspects

cosmic elemental forces and particles 67, 69, 82, 83, 84
amygdala 95
anathema
  archetypal 137
  secular 115
anatomy
  wired/ physical 98
  wireless 98, 103, 107
animistic language/ animism 144
anti-entropy .... *see* negentropy
anti-gravity .... *see* levity
anti-parallels
  (*raj*/ *tam*) cosmic fundamentals .24, 86
  cosmic/ dialectical stack vectors 24
  informative/ energetic dipoles ..56, 61, 138
  interpretations
    of biology 138, 140
    of physics 118
  inversion ...... 24, 60, *see* also inversion/ reflective asymmetry
  knowledge inward/ outward action.85, 86
apoptosis/ programmed cell death ...... *see* death
appearance .... *see* illusion/ reality
archetype
  > cosmic grammar 69, 82
  anathema 137
  archetypal memory . 13, 42, 83, 94, 97, 100, 116, 117, 118, 119, 121, 123
  archetypal program. 96, 100, 102, 137, 138, 139
  archetypal projection ..... *see* creativity/ reason/ creation
  bio-definitions ... *see* bio-classification
  bio-logical 100, 107, 137, 138, 140
  discrete template/ theme/ type/ invariant principle 107, 112, 114, 138
  harmonic..106, 107, *see* also harmonic oscillation/ vibration/ resonance/ music
  hologram/ holographic transmission 101, 115, 117
  immanence/ potential/ latency .... *see* potential
  informative potential 137, 138
  informative source/ metaphysical egg/ cosmic reason 112, 117, 121
  intrinsic/ space-time embedded quality 115
  mathematical abstraction 116
  metaphysical control 101

natural code 83
natural reason/ cause/ law... 73, 74, 77, 118, 121
psycho-biological ... 101, 107, 138, *see also* bio-logic/ potential/ polarity
psychological/ metaphysical entity 115
subconscious form/ file in universal mind... 95
typical mnemone ...*see* mnemone/ bio-logic/ potential

Archetype
Cosmic First Cause 119
First Cause Psychological 57, 119, 123
Logical 57, 118
Primal Motion .... 119, *see also Logos*
Aristotle 54
as if...................................*see* metaphor
asex ....*see* reproduction
atom............ *see also* sub-atomic particle
atoms-to-man?...................... 114, 142
atoms-understanding-atoms?........... 87
collapse of wave function.............. 114
enduring structure.......................... 122
harmonic oscillator................ 106, 107
non-conscious/ oblivious .. 10, 147, *see* oblivion/ non-consciousness
nucleus/ intra-atomic influence 125
perpetual motion............................ 147
Planck's view 128
symbol of physical order/ chord/ word 84, 125

attractor
attractor = archetype........................ 56
flexible psychological > desire .... 92, 137, *see also* psycho-logic/ volitio-attractive force
inflexible > first cause physical.... 115, 117, 119, 137
magnetic influence 137
morphological attractor/ morphogene .... morphogene/ typical mnemone
pull-to-future/ behavioural constraint 56, 115, 137
sub-conscious record ....57, *see* archetypal memory/ potential matter/ mnemone

Attractor
pull-to-Super-Consciousness..........*see* Archetype/ *Logos*
AVB/ Authorised Version of Biology 142
axiom 9, 12, 13, 33, 146
Axis....................... *see* Central Essence
axis/ centre-line ..... 34, *see* source/ pivot/ point of balance

B

balance ... *see also* neutrality/ symmetry/ equilibrium/ homeostasis/ health
balanced imbalance/ conservation law ...... 117, *see also* equilibration/ law
equal-and-opposite effects cancel/ equation satisfied/ resolution 117
harmonic counter-point 112, 122
imbalance > motion/ change.......... 122
inertial equilibrium/ slack/ impotence. 61
informative equilibration/ balanced orchestration/ homeostasis ... 18, 19, 20, 23, 24, 32, 108, 114, 132
peace/ resolution.............................. 62
pivotal 14, 18, 20, 21, 24, *see also* axis/ centre-point/ zero-point/ dot
pre-dynamic potential/ poise 16, 19, 44, 62, 88, 91, 117
psychological well-being/ basis of mental health.............................. 92
Sat cosmic fundamental/ quality/ axis/ hub/ core .. 21, 23, *see also* essence/ centre/ source
social equilibrium 150
being...................................*see* essence
Best Criterion 88, 147, 149, 152, 154
big bang................ *see also* transcendent projection/ miracle/ zero-point
illogically, from absolutely nothing. 45
inflation 45

bio-classification
archetypal/ typological level. 100, 101, 138
archetype
discrete template 101, 137
natural unit of bio-logical program 98, 100, 101
evolutionary continuity/ phylogeny 141
holistic/ dialectical classification 96, 98, 100
natural unit of bio-logical program 137
X. archetypalis..............................100
Biogenetic Law (Pasteur) ...... *see* Law of Biogenesis

bio-logic
bio-logical explosion/ code-generating big-bang.................................... 143
low-level, local variation within high-level, general invariance ....*see* variation/ principle
reason-in-biology................... 108, 137

bio-logical development
adult form/ end-frame.....................*see* reproduction/ egg
archetype > physical expression ... 137, 138, 139
hierarchical procedure 138
hierarchically informed control 138

informant program > informed body 137, 138
informative seed > adult form 132
inversion/ inside information outwardly expressed 60, 61
symbolic code > physical expression 138
bio-rationality/ core bio-rationality ....*see* code/ language/ bio-logic/ informative potential/ genetic grammar & linguistic hierarchy/ super-code
bio-reason .... *see* bio-logic
Bischof Marc 102
black box
   evolutionary 88, 94, 95
   material 122
   metaphysical/ psychological..... 88, 95, 137
black hole ... *see* also apparent absolutes/ extreme subtendence/ death/ pain/sin/ immorality/ negative power
   material oblivion .... *see* non-consciousness
   negative extreme 61
   physical 50
   psychological .... *see* sleep/ sub-consciousness/ negative power
Bohm David 114
Bohr Niels 58, 117, 128, 146, 149
Born Max 114
Bose Jagdish..................................... 128
Bose Satyendra................................. 128
bottom-up 12
brain
   mind = brain .... *see* consciousness/ materialism/ neuro-hormonal system/ reality
   neural oscillation/ brainwaves 94
   psycho-somatic mediator................. 87
   structures 110
Buddha 77, 147, 153
Burr Harold 102, 103

C

caduceus
   logical/ conceptual human program 100
Cambrian explosion . *see* fossil/ bio-logic
cause
   can cause itself < boot-strap logic ... 55
   can't cause itself 35, 54, 55
   causal mind ....*see* (higher) Archetype/ *Logos*/ (lower) archetype/ mnemone
   conceptual-teleologic pull <> energetic push 56
   quantum causation/ matter-in-principle > matter-in-practice..... 28, 35, 67, 69, 73, 101, 107, 123
   sufficient/ principle of causality 54
   teleological code > bio-logical effect 137
   'transcends' effect 125
Cause .... *see* First Cause
cell
   abiogenetic impossibility............... 108
   basic (electromagnetic) unit of life .. 48
   bio-electrics 97, 102, 103, 107
   codified chemistry 67, 135
   collection of very specific chemicals 10
   dormant/ sub-conscious mind or typical mnemone in every cell .100, 101, 102
   information incarnate 135
   information primary/ material arrangement secondary 131
   intra- and inter-cellular information systems 138
   mind machine 137
   psychosomatic aerial 100, 103, 106
   soft musical machine 48, 106
   spontaneous generation 144
Central Dialectic....*see* Natural Dialectic of Polarity and dialectical factor/ Primary Dialectic
centre/ centrality .... 23, *see* source/ pivot/ axis/ point of balance/ zero-point
chance
   accident/ coincidence/ contingency .64
   code = anti-chance............. 78, 81, 137
   HUP (quantum uncertainty principle) 114
   Lady Luck 78, 133, 141
   luck/ unexpected, unpredictable event 133
   mind = chance-reducer/ anti-chance 81
   mindless/ passive/ unintelligent 'creator' 78, 79
   Penrose computation > cosmos-not-by-chance 114
   unreasonable/ incalculable/ untestable 108
chaos .... *see* also chance/ randomness/ dysfunctional logic
   pattern-less confusion............. 92, 153
chemistry......................................70, 84
Chladni Ernst................................... 104
Christ.................................. 77, 91, 153
circuitry
   computer chip/ integrated circuit .. 136, 139
   cybernetic/ bio-systematic .... 65, 132, 137, *see* also homeostasis

electronic 105, 150
genetic circuitry ...... 135, 139, 143, *see
also DNA/* computation
nervous 95
code..*see* also information/ computation/
purpose/ *DNA* molecule/ alphabet/
bio- & cosmo-logical languages
cosmic .... *see* cosmic language
genetic/ bio-logical 101, 132, 136
idea registry .... 69, 79, 100, *see* also
memory/ *DNA*/ computation/
language/ archetype
informative potential 84, 136
is metaphysical 67, 100
language < mind .... *see* information/
innovation/ informative
negentropy/ source
languages bio- and cosmo-logical ..*see*
alphabet/ quantum aspects/ cosmo-
logical language/ vibration/ genetic
grammar
meaningful/ grammatical structure . 81,
83, 131, 133
program/ teleological design/
algorithm ... 49, 65, 68, 81, 96, 118,
136, *see* code/ bio-logic/ nucleic
acid operations/ archetype
signal organiser 81
specific/ purposeful information 69,
83, 108, 133
top-down programming > main/
master routine 112
top-down programming > sub-routine
.... 100, 101
complexity non-purposive .... 77, *see* also
nature/ law/ archetype/ passive
information
complexity purposive .... 78, 108, *see* also
order/ innovation/ creativity/ active
information
computation
bio-operating system 134, 135, 139
computer = mind-machine 65, 135
modular programming 138
top-down programming > main/
master routine 112, 136, 138
top-down programming > sub-routine
.... 112, 118, 136, 138
Turing machine 135
concentration .... *see* also focus/ source
biological/ concentrated information/
code/ high negentropy/ order 130
physical/ energetic-massive 23
Concentration
Pure Concentrate of Consciousness 23,
37, 59, 91, 147, 154, *see*
Consciousness

conceptual development
act of creation 77
anticipatory procedure/ codified
algorithm > goal 82, 143
archetype > physical expression 137
conceptual approach to biology 132,
138, *see* also bio-logic/ archetype
fundamental order of 67, 77, 78, 79,
see also hierarchy
immaterial/ conceptual/ metaphysical
> material/ factual/ physical 122,
137
inversion/ inside information
outwardly expressed 60, 79
conscio-material (c-m) dipole
anti-parallel vectors 24, 33
basic existential dipole 73
c-e conscio-energetic spectrum. 15, 52,
75, 97
c-m gradient/ spectrum/ slope/ scale/
chiaroscuro .. 15, 18, 24, 62, 75, 119
cosmos/ creation as dipole 14, 23, 57,
59, 60, 147
informative/ immaterial > energetic/
material coupling 33, 65, 73, 79, 83,
85, 88, 108
informative/ subjective/ psychological
> energetic/ objective/ physical
coordinates 62, 73, 85, 88
sliding scale/ proportional
representation 15, 33
source-to-sink 15, 16, 62, 149
conscio-material gradient 43
consciousness .... *see* also metaphysic/
active information/ active mind
altered states of (*ASC*) 78
basis of awareness/ experience . 13, 22,
40, 72, 87, 88, 92, 147, *see* also
knowledge
evolution of your *sine qua non*? 88
illusion suffering an illusion 87, 147
immaterial element/ metaphysical
entity 9, 13, 52, 61
intelligence .... *see* intelligence
materialistic view/ illusion suffering
an illusion/ neuro-scientific
reductionism .. 10, 87, 97, 146, 147,
see also illusion/ reality/ matter/
neuro-hormonal system
mind = ever-changing/ formful flux 68,
80
qualities/ levels 76, 90
seamless perception/ unitary
coordinator 128
Consciousness .... *see* also Essence/
Absolute
Communion 53, 77

Cosmic Axis 88
Dharmakaya/ Nirvana/ Samadhi/ Moksha/ Radiant Communion/ Transcendent Bliss 88, 91, 148, 154
Essence/ Absolute Truth 147
Experience of Pure Consciousness . 52, 53, 154
Highest Value/ Ultimate Criterion .. 88, 151, 154
Illumination/ Enlightenment 40, 55, 59, 88, 91, 151, 153, 154
informative potential/ latent mind .. 22, 40, 59, 92, 147, *see also* source/ potential/ information
Natural Substance 88
Primary Component of Creation 88, 128
Psychological Principal/ Supreme Being/ Quintessence 88
Pure Concentrate of Consciousness/ Life/ Soul 40, 52, 59, 61, 76, 88, 91, 147
Source of Informative Mind 88
Superintendent Pole *opp.* base-pole's oblivion 88, 91
Super-State 48, 75, 88, 91
Uncreated/ Uncaused 55, 88
consciousness-in-motion .... *see* mind
conservation
   natural law 117
   of bio-modular sub-routines ... 139, *see also* homology/ code/ archetype
constant .... *see* physical constants
contemplation
   inward concentration of attention ... 37, 92, 147
   metaphysical connection 120
   opp. sensation 90
convergent evolution ... 138, 139, *see also* Glossary/ computation/ code/ modular programming/ adaptive potential/ homology/ mosaic distribution
cosmic fundamentals *see also* dialectical operator
   dialectical operator 21, 27
   fundamental order of creation .. 18, 19, 22, 30, 34, 61, 65, 73, 76, 79, 85, 119, see also conscio-material dipole/ cosmic models/ hierarchy/ source > sink
   physical expression/ link to physics 18, 22, 30, 61, 115, 125, *see* archetype/ potential (matter)/ quantum aspects/ matter
   psychological expression/ link to psychology 22, 30, 76, 90, 115
   raj upward/ active/ levitational vector 21
   Sat pivotal/ neutral/ equilibrator 21
   tam downward/ passive/ gravitational vector 21
   trinity/ triplex nature ... 21, 22, *see also* hierarchy/ Natural Dialectic
cosmic models
   (clockwork) machine 105, 128, 150
   concentric rings/ vibrations/ waves. 16, 17
   conscio-material (c-m) gradient. 16, 62
   Cosmic Pyramid 15
   four-square pyramid/ ziggurat .. 16, 34, 72, 75, 86
   hologram 117
   Master Model/ Central Metaphor/ Music 104, 105
   mind-machine/ computer 128
   Mount Universe 16, 21, 40, 72
   scale/ balance 19, 24, 62
   spectrum 15, 16, 97
cosmo-logic 76, 114, 117
   fundamental order of creation. *see also* order/ cosmic fundamentals/ conscio-material dipole
   fundamental order of physics ... 22, 30, 61, 120, 124, *see also* nature/ order/ cosmic fundamentals/ trinity
   fundamental order of psychology ... 22, 30, *see also* cosmic fundamentals/ hierarchy/ trinity
   inversion/ inside information outwardly expressed ...... 60, 79, 83, 114, 119, see also conscio-material dipole/ anti-parallels/ order/ conceptual development
   low-level, local variation within high-level, general invariance .... *see* variation/ principle
cosmo-logical axis .... 36, 154, *see also* cosmos/ eye-centre/ point X
Cosmo-Logical Axis (Universal) ...... 37, 151, *see also* Pivot/ Zero-Point/ Central Essence
cosmo-logical language/ score
   atom/ chord/ word 84, 125
   forces/ textual (re-)arrangement 125
   language of physics & chemistry 83, 125
   meaningful/ grammatical structure .. 67
   molecule/ phrase 125
   particle notes > atomic chords > molecular phrases 67, 84, 125
   sub-atomic particle/ note/ letter .... 125, *see also* quantum aspects
cosmos .... *see also* creation/ conscio-material dipole/ universe/ polarity/

cosmic fundamentals
   archetypal pattern 55, 114, 115
   can't create itself 55
   cosmic program/ variation-on-
      archetypal-theme .... *see* cosmic
      language/ quantum aspects/ fine-
      tuning
   cosmogony/ transcendent projection
      .... *see also* big bang/ source/
      transcendence
   cosmos < vibratory sustenance ....*see*
      vibration/ energy/ perpetual
      motion/ harmonic oscillataion/
      appearance
   dynamic/ vibratory text 53, 82, 104,
      128
   hierarchical 66, 72
   information-carrier .... 73, *see also*
      informative potential/ nature/
      matter/ archetype/ law
   macrocosm/ cosmos 119, 154
   microcosm/ man 85
   orderly/ non-chaotic 73, 79, 112
   perpetually imbalanced balance/
      perfect imbalance .... see motion/
      balance/ equilibration
   projected. 36, 38, 40, 45, 55, 111, 114,
      115, 121, 128, 148
   source 50, 77
   trinity/ triplex nature..... 14, 22, 32, 33,
      40, 60, 72, 85
   universal body 125
creation/ act and fact .. *see* also conscio-
   material dipole/ fundamental order of
   creation/ cosmos/ innovation/
   creativity
   bio-logical/ archetypal projection .. 137
   cosmo-logical projection ... *see* cosmo-
      logic/ cosmo-logical language
   counter-creation/ destruction 77
   created object/ form/ event .. 22, 62, 69
   evolutionary mechanism.. 13, 140, 142
   expression of *tam* vector/
      materialisation ....44, 48, see also
      impotence/ energy
   field of relativity 20, 32, 114
   hierarchical order of materialisation
      ...... 16, 33, 61, 68, 72, 79, 118, 124,
      132, 140
   inversion. 60, *see* symmetry/ reflective
      asymmetry
   macrocosmic/ universal .... *see* also
      transcendent projection/ archetypal
      expression/ conscio-material
      dipole/ big bang
   microcosmic/ individual instance
      mindful 23, 77, 79
      mindless ... 42, 49, 121, *see* passive
      information/ reflex cause/
      nature/ non-consciousness
   object of psychological or physical
      expression 37
   opposite vector/ perception/
      comprehension/ knowledge 77
   orderly constraint/ reduction of
      freedom 113
   perpetual motion/ change . *see* motion/
      energy
creativity
   active information .... *see* also
      information/ mind
   hierarchical order of 65, 75, 77, 78, 79
   idea/ source of invention 11, 22, 92,
      138
   matter is creatively impotent 65
   subjective (re-)arrangement of
      information 22
creator
   informative source 148
   Logos/ Sound and Light 53
cymatics 104

D

Darwin Charles 142, 143
Darwin Erasmus 145
Darwinian puddle .... 142, *see* also
   abiogenesis
data item 84
death
   codified self-sacrifice/ apoptosis ... 135
   end. 11, 44, 50, 86, 125, 133, 134, 135
delusion
   code-specifying matter 139, 143
   law of non-conscious innovation ... 143
   material anticipation or conceptual
      teleology .... 139, 143, *see* also
      anticipation/ matter/ teleology
   molecular self-animation 142
   special delusions .... *see* reality
dematerialisation/ dissolution
   (*raj*) upward/ levitational vector (↑) 76
   expression of (*raj* ↑) upward/
      levitational vector 24, 76
   inward/ raising process 76
depolarisation (two > one) *see* neutrality/
   equilibrium/ unity/ love
desire .... *see* psychological force
dialectical factor
   element 25
   member 25
   method 7, 12
   operator . *see* also cosmic fundamental
   tam/ *yin* downward/ passive/
      gravitational 21

operator *raj/ yang* upward/ active/
    levitational 21
operator *Sat/ Tao* pivotal/ neutral/
    equilibrator 21
Primary Dialectic 27, 28
Primary Stack 28
secondary stack 29
stack 25
tri-logical stack 26, 27
vectors 27
differentiation ... *see* variation/ (dis-)unity
DNA molecule
    as microchip/ rigid memory 68, 101,
        see also circuitry
    chemical information carrier/ paper-
        and-ink 68, 84, 132
    database/ genome/ information store
        .... 68, 102, 135
    local physical expression of general
        metaphysical archetype 67, 100,
        132
    physical information storage 136
    secret of life? 91
    solenoid/ aerial? 100, 103
    symbolic bio-code/ vital text 68, 69,
        82, 100, 132, 135
Dobhzhansky Theodosius 130
dormant mind .... *see* sub-consciousness
dot 24
dysfunctional logic
    all is matter (materialism = holism). 99

E

Eddington Arthur 116
egg
    potential/ capability 101
ego
    intellectual executive 92
Einstein Albert .. 112, 116, 117, 128, 177
electrical charge 61, 97, 98, 103, 116
electromagnetism .... *see* force physical
energy
    > existence 147
    archetypal behaviour .... *see* law/
        potential/ archetype
    energetic coordinate/ secondary
        component of existence 62
    energetic trinity .... *see* cosmic
        fundamentals/ physic's primal
        trinity
    non-conscious/ objective .... *see*
        objectivity/ non-consciousness/
        matter/ teleology
    protean basis of physics 44
    psychological .*see* force psychological
    whence came/ comes energy? 45
engram 95

entropy ..... *see* also energy/ information/
    sink/ sub-state/ gravity/ Glossary
energetic
    exhaust from system/ run-down/
        drag 56
    expression of physically dominant
        (*tam* ↓) vector 56
    fall > sub-state sink/ gravitational
        vector of recession 93
informative
    fall to sub-conscious sink 93
    randomising/ diffusive tendency 56
enzyme .... *see* metabolism/ protein
equilibration .... *see* balance/ *karma*/
    equilibrium
equilibrium
equilibrium dynamic
    cyclic action/ wave/ vibration 24,
        132
    equilibrium inertial
        field of no possibility/ exhausted
            end-product 24, 61, 93
equilibrium potential .... *see* source/
    poise/ (*sat*) cosmic fundamental
    field of possibilities 16, 24
essence
    lesser essence/ relative or existential
        being ...*see* existence/ individuality/
        (dis-)unity
Essence
    Archetype 119
    Cosmic Pivot/ Centre/ Axis 20
    Cosmic Potential 23, 40, 45, 55
    opp. existence 28, 57, 62
    paradoxically includes existence 27,
        28, 40, 54, 59
    Supreme Being 19, 32
    Truth Absolute 37, 147
    Ultimate Reality 37
Essential Dialectic .... *see* dialectical
    factor:Primary Stack
evolution
    Darwinian theory 9
evolutionary theory
    chemical evolution ..... *see* abiogenesis
    Darwinian/ chance and natural
        selection .... *see* also Glossary
    evolution of natural law? 145
    materialism's *sine qua non* 9
    neo-Darwinian/ synthetic. 13, 139, 142
exhaustion .. *see* also energy/ impotence/
    nothingness/ sink
    expression of (*tam*) cosmic
        fundamental 17
    opp. potential 17
existence ...... *see* also Glossary/ polarity/
    motion/ conscio-material dipole/

relative illusions
  appearance < motion 147
  basic coordinates/ coefficients 147
  compound of mind and matter ... 54, 55
  conscio-material field/ psycho-
    physical nature 62, 147
  creation-is-motion 32
  di-polar 14, 32, 73
  Essence-caused/ Essence-dependent/
    Essence-in-action 32, 45, 48, 54
  existence < vibratory sustenance .... *see*
    vibration/ energy/ motion/
    harmonic oscillataion/ appearance
  hierarchical order .... 40, 100, *see*
    hierarchy/ order/ archetype/ cosmo-
    logic
  metaphysical non-existence/ physical
    presence 147
  opp. Essence/ non-Essence/ non-
    essential pole 32, 54, 57
  physical non-existence/ metaphysical
    abstraction 147
  projection/ relative illusions 147
  variation-on-archetypal-motions .... *see*
    motion/ energy/ archetype/ cosmos/
    appearance/ cosmo-logical
    language
  vectored 19
  zone of relativity 32, 50, 54
existential dialectic .... *see* **dialectical factor**, *see* secondary dialectic
existential dipole 63
experience
  basis = consciousness 88
  brain mediates physical experience 87,
    see also mind/ brain/ neuro-
    hormonal system
  graded/ sliding-scale/ hierarchical
    priorities 149, 151
  living being 87
  made of molecules-in-nerves? . 87, 150
  sole/ subjective form of knowledge 24,
    40, 88, 147
  sole/ subjective form of understanding
    .... 24, 92, 118, 153
Experience .... *see* Supreme Being/
  Consciousness
explicit universe 115
eye-centre 37, 154, *see also* cosmological
  axis/ third eye/ concentrative focal
  point

F

faith
  animistic 144
  materialistic/ naturalistic 139
  promissory 146

scientism 87, 149, 151
secular 153
two pillars of 151
Feynman Richard 123
fine-tuning
  biological 139, 143
  cosmic 114
  cosmic music/ soundless harmony 114,
    see also vibration/ harmonic
    oscillation/ cosmic models
  forces and particles articulate .... *see*
    cosmo-logical language/ resonance
first cause physical
  bio-logical .... 137, *see* creation/
    archetypal projection
  physical ..... 55, 118, 119, 120, *see* also
    archetype/ potential matter / big-
    bang/ transcendent projection/
    Glossary
  psycho-biological 100, 137
First Cause Psychological
  Archetypal .... *see* Transcendent
    Projection/ Logical Archetype/
    Logos
  Cosmic .... 52, 55, 148, *see also Logos*/
    Glossary
  Cosmic Egg (is metaphysical) 23
  cosmos not self-caused 73
  Natural Scientific Cause 55
  Primal Motion 55, 57
  Psycho-logical ... 85, 118, 119, *see* also
    Logos/ Logical Archetype
  Uncaused 55
First Principle 92, 151
focus .... *see also* information/
  concentration/ centralisation/
  meditation
  concentration of attention .. 78, 80, 147
  concentration of information .. 130, *see*
    also principle/ informative
    potential/ law
force
  electromagnetic 109
  physical
    cosmic operators 49, 56, 84, 99,
      105
    whence?*see also* energy/ motion/
      never codifies 145
      whence? 112
  psychological .. *see fig.* 12.1 also mind/
    desire/ passion/ love
fossil 143, 145
freedom
  free will and determinism
    conditioned free will/ degrees of
      choice 92
  relative constraint 113

frozen time ....*see* sub-consciousness
functional logic 54
fundamental order of creation *see* cosmic fundamentals/ conscio-material dipole/ nature/ order

G

gene
  jumping gene/ retrotransposon 136
  jumping gene/ transposon 136
  micro-hierarchy/ genetic regulation/ expression 136, 138, 139
genetic grammar
  binary/ digital code 135
  epigenetic markers/ punctuation 143
  syntax 69, 82
genetic linguistic hierarchy
  genome/ encyclopaedia/ book of life
  .... 100, 101, 137
genome ... *see* genetic linguistic hierarchy
glass ceiling 126
godless gap 143, 145
governing template .... *see* archetype
gravity/ anti-levity ...... *see also* polarity/ cosmic fundamentals/ anti-parallels/ force physical/ materialisation/ Glossary
  dialectical sense
    cosmic influence of (*tam*) descendent vector .... *see* also cosmic fundamentals/ negative power
    dark/ physically dominant force . 24
    drag/ drain/ resistance/ sink/ inertial influence .... 24, *see* sink/ matter/ inertial equilibrium
    energetic loss > exhaustion/ finished state .... *see* entropy
    Higgs mechanism ...... *see* also mass
    loss of concentration > dispersal/ weakening .... *see* diffusion/ variation/ randomness
    subjective - informative loss > noise/ nonsense/ non-conscious state .... *see* informative entropy/ psycho-logic/ randomness/ oblivion
    subjective > inc. suffering/ igorance/ unhappiness/ pain . 76, 92, *see* **also morality/ evil**
    tendency to compress/ contract/ bind/ aggregate . 76, 92, 123, *see* also materialisation
  scientific (and dialectical) sense
    mutual attraction of physical bodies/ gravitation/ gravitation
    56

GTE (general theory of evolution) ......*see* macro-evolution/ transformism/ innovation/ type/ archetype

H

H. archetypalis
  human program 100
  memory man 100
  psychosomatic form/ mind-side 101, 104
  subconscious/ psychological side . 102, 104, 109
H. electromagneticus
  electrodynamics 97, 101, 102, 104
  non-conscious bioelectrical aspect .. 97, 103, 109
  wireless/ radiant anatomy 98
H. sapiens
  classical/ biological form 98, 99
  molecular and visible body 103
  sub-state shell 98
  wired/ physiological anatomy 98
habit *see* sub-consciousness and personal mnemone
harmonic oscillation ... 103, 104, 107, *see* also resonance/ vibration/ music/ atom
health
  archetypal/ psychosomatic resonance
  107
  homeostatic stability 134
  pre-set normality 107
  vis medicatrix naturae 107
Heisenberg Werner 128
helix .... *see DNA* molecule/ caduceus
hierarchy
  bio-logical hierarchies ... 132, 134, 138
  computational hierarchies ...... 136, 139, 143, *see* also computation/ informative hierarchy/ code
  dialectical 16, 22, 30, 61, 112, 125, *see* also fundamental order of creation/ cosmo-logic/ cosmic fundamentals/ Natural Dialectic of Polarity
  informative > energetic act of creation ... 49, 65, 76, 79, 112, 124, 136, *see* also conceptual development/ creativity/ information
  models .... *see* cosmic models of truth/ reality 147
  psycho-biological 140
  symbolic 16, 143
  triplex/ 3-tiered
    general/ cosmic . 16, 32, 33, 40, 66, 72, 124
    sub-division physical 22, 30, 35, 61, 72, 112, 124

sub-division psychological .. 22, 30, 34, 76, 90, *see* also mind/ experience/ subjectivity
hippocampus 95
holism 10, 74, 115
both immaterial-material/ psychological-physical/ informative-energetic components of cosmos included 11, 40, 147
dialectical expression 14
hologram .... *see* archetype/ cosmos
holy grails
GUE psychological/ metaphysical - Grand Unified Experience/ Highest Goal/ Holy Grail/ Communion .. 77, 91
psychosomatic/ mind-body connective mechanism .... *see* psychosomasis
homeostasis
bio-systematic 132, 134
dynamic/ developmental norm 134
negative feedback 134
norm/ keeping the balance/ cyclical equilibrium 134, 135
triplex mechanism/ sensor/ regulator/ effector 132, 134
homology *see* also bio-classification/ tree of life/ computation/ code/ modular programming
archetypal routine/ bio-modular programming 138
mosaic distribution 139

I

idea .. *see fig.* 12.2 - The Order of an Act of Creation
rationalisation of possibility .... *see* reason/ principle/ teleology
registration of possibility ...... *see* code/ memory
Idea. *see* First Cause/ Metaphysical Egg/ Top Teleology
ignorance *opp.* knowledge 93
illumination
knowledge/ grasp of meaning/ wisdom 24, 55, 92, 153
Illumination .... *see* Consciousness
illusion .... *see* also delusion/ reality/ Glossary
Buddhist view 147
existential appearances = relative/ lesser realities 147, 149
geno-centric vision 149
macro-evolutionary transformism .. *see* macro-evolution/ unlimited plasticity
oblivious design 139
immaterial element *see* also information

information 64
mind 64
immortality
eternal matter? 45, 67, 127
logical/ mathematical/ metaphysical principles 117
Transcendent Super-State 91
Uncaused Cause/ Pure Deathless Life 54
implicit universe .. 115, *see* also universal mind/ archetype/ potential matter
impotence .... *opp.* potential *see* also exhaustion/ subtendence/ sub-state
expression of (*tam*) cosmic fundamental *see* also materialisation
final cosmic phase/ non-conscious energetic matter 69, 83, 88
no more possibility/ exhaustion/ end-point 24, 44, 61
incarnation
biological organism 96
individuality .... *see* (dis-)unity/ ego/ separation
Infinity ...*see* Absolute/ Natural Essence/ Trancendence
Cosmic Potential 52, 126
Metaphysical/ Psychological Absolute 147
inflation .... *see* big bang/ levity
Informant, The .... *see* Logos
information
active/ passive .. *see* also information - dialectical sense of mode
bio-information 74
bio-explosion/ code-generating Cambrian big-bang 143
bio-information 130, 135
cell metaphysical/ seen as information 131
informative coordinate/ primary component of biology. *see* code/ bio-logical language/ genetic grammar/ genetic linguistic hierarchy
life-form = information incarnate 130, 131, 132
specific, complex information ... 74, 143
cosmic element
cosmic pole 59, 69, 147
immaterial/ metaphysical/ psychological 56, 64, 65, 74, 88, 99, 115, 140
not = energy/ not from matter ... 65, 73, ***see* also matter/ innovation/ creativity**

186

dialectical sense of balance
'down'............ *see* gravity/ entropy/ information loss
'rest'...................... *see* equilibrium/ centralisation/ understanding
'up' ....*see* levity/ negentropy/ information gain
dialectical sense of direction
'outward/ entropic'....... 56, 77, 153
dialectical sense of mode
active/ conscious 65, 68, 73, 79, 80, 83, 100, 119, 150
passive/ archetypal medium/ first cause physical 97, 102, 120, 138
passive/ unconscious 48, 65, 68, 69, 73, 77, 96, 119
hierarchy
> order > logic 79, 84, *see* also law/ hierarchy/ fundamental order of creation
conscio-material scale .. 33, 65, 148
informative hierarchy.... 33, 76, 90, 132, 134, 137, 140, 143
information technology 135
informative density
DNA.. 143
of principle. 100, 119, 124, *see* also law/ principle/ morality
informative entropy........*see* mutation/ noise/ randomness
informative transmission/ exchange command & control/ central processor 22, 64, 85, 102
psycho-biological.... 49, 55, 88, 98, 103, 107
informative trinity .... *see* Super-Consciousness/ consciousness/ unconsciousness
material/ scientific view - information as an aspect of automatic energy 64, 97
metaphysical ingredient.................... 10
passive.....*see* information - dialectical sense of mode
source > outcome (sink)
causal information/ precedent trigger/ behavioural guide..... 24, 48, 53, 80, 137, 150
informative potential/ precedent latency 17, 73, 82, 83, 112, 119, 132, 138
physical expression/ bodily behaviours ....*see* physical phenomena/ bio-logical development/ bio-logical forms
physical storage *see DNA* molecule/ database/ other physical storage media
source of information ... 42, 82, 130
storage metaphysical 48, 95, 117
informative potential ..... *see* information/ potential/ code/ archetype/ consciousness/ cosmos/ law/ nothingness
infra-conscious = sub-conscious........... 52
Initial Projection.................... *see Logos*
innovation 78
creativity........ 11, 53, 75, 79, 130, 143
imaginary law of non-conscious innovation 145
matter can't create code .. 81, 139, 143, 145
rapid 143
instinct
archetypal behaviour 42, 48, 68, 90, 119, 121
behavioural reflex........................... 77
mnemonic file/ sub-routine..... 49, 101, 103, 132, 137, 138
natural (bio-)law 82, 150, 154
sub-conscious/ mnemonic mechanism 48, 57, 103
sub-rational/ involuntary/ thoughtless 92
intelligence
grasp of principle...................... 16, 73
message/ relevant data.................... 73
metaphysical power/ acumen........... 67
unconscious intelligence = passive information ... 133, *see* also instinct/ program/ computation/ machine/ creation (created thing)/ physical phenomenon
inversion .... 15, 24, 46, 48, 60, 61, *see* creation/ conscio-material dipole/ symmetry - reflective asymmetry/ anti-parallels/ biological development/ switch
diabolical dialectic........................ 152
irreducible complexity = irreducible organisation 114, 134

J

Jeans James 128
Jenny Hans 104

K

Kalam-i-Illahi/ Kalam/ Kalima ....*see Logos*/ Sound and Light
karma
action/ reaction 20, 56
action-in-the-cosmic-balance/ equal-and-opposite-effect 56

fields of action (psychological and physical) 20
knowledge .... *see also* understanding/ illumination
   active/ seeking answers ... 92, 113, 128
   passive/ rote..................................... 92
   understanding explicit physical cause > cleverness 24
   understanding implicit psychological cause > wisdom 24, 55, 153

L

language
   descriptive/ instructive 81
   genetic/ bio-informative 83
   grammatical/ syntactical structure .. 69, 82, 130, 135
   symbolic information 81, 130, 135
   universal (music and mathematics) 104
   vehicle of communication/ meaning .... 69, 81, 102, 135, *see also* information
Lao Tzu 38
law
   bio-logical .... *see* bio-logic/ archetype/ mnemone
   external/ ethical canon & governmental law 150, 152
   internal/ moral code........ 150, *see also* morality
   latent instruction/ informative potential 73, 78, 111, 115
   machines accede to physical law 84
   metaphysical/ archetypal 100, 105, 117, 119
   natural law = eternal 'necessity' 78, 79
   natural law/ natural behaviour/ physical reflex ... 13, 25, 48, 65, 73, 121, *see also* archetype/ cosmo-logic/ bio-logic/ instinct
   natural psycho-reflex/ instinct 82
   relative moralities/ rules/ ideologies 154
   resonant enforcement 105
   rule and regulation......................... 112
   symbolic/ grammatical 69, 82, *see also* cosmo-logical language/ score, *see also* cosmo-logical language/ score
Law
   Natural/ Essential/ Absolute .... *see* Communion
Law of Biogenesis............................ 142
lesser truth/ reality *see* appearance/ truth/ reality/ conscio-material grading from Reality
levity/ anti-gravity ...... *see also* polarity/ cosmic fundamentals/ anti-parallels/ force physical/ dematerialisation/ Glossary
dialectical sense
   cosmic influence of (*raj* ↑) ascendant vector ..... 24, 123, *see also* cosmic fundamentals/ positive power
   lift/ dynamic influence 52, 123, 153
   subjective - informative gain > accurate message/ sense/ conscious state .. *see* informative negentropy/ psycho-logic/ order/ teleology
   subjective - spiritual ascent . 91, *see also* concentration/ centralisation/ Transcendence
   subjective > inc. understanding/ happiness/ well-being/ love .. 76, 92, 153
   subjective-informative/ objective-energetic stimulus.......... 55, 123
   tendency to decompress/ expand/ unbind/ release .. 24, 76, 92, 123, 153, *see also* dematerialisation/ dissolution
scientific (and dialectical) sense
   vacuum energy *see* space/ quantum aspects/ ZPE
life ... *see also* consciousness/ experience/ essence
   holistic inclusion
   biological form = embodied mind 99, 137
   holistic definition .... 40, 54, 61, 85, 91, 154, *see also* Essence/ Soul/ Consciousness
   life-form = information incarnate 84, 98, 130, 136
   metaphysical informant................ 9
   mind/ consciousness/ immaterial element 80
   opp. death/ matter/ non-life 9, 49, 147
   vectored (ascent/ balance/ descent) 152
life's highly specified/ codified bio-complexity 78, 84, 108, 136
life-fit/ friendly universe................ 114
life-style................................. 150, 154
materialistic reductionism
   accidental atomic configuration? 87, 114, 142
   chanced from a pennyworth of chemicals? 9, 142
   electronic after-thought/ an experience of nervous atoms? .... 87, *see* neuroscience/ reality
   reductionist definition 91

light
information-carrier 102
life-giver 102
physical
　apparent absolutes 51
　least material levitational entity . 35
　pure radiant energy 35, 49
psychological
　guilt-less state 65
　Light of the World/ Logical
　　Illumination 53, 75
　volitio-attractive push-pull current
　　.. 80
psychosomatic/ psycho-biological
　'body of light' .... see *H.*
　　electromagneticus
medium/ connector 98, 104
Lister Joseph 144
logic
algorithmic 119
deductive 116, 119
Logos .... *see* also Causal Information/
　Psychological Potential/ Creator
First Cause/ Primal Motion/ Initial
　Projection ...... 79, 118, 149, *see* also
　Creator
Primary Logic 69, 83
Source 16
Source of Mind/ Psychological
　Substance 88
Vibratory Communicant 53
love
compassion/ radiant beneficence/
　blessing ..... 148, 153, *see* also saint/
　holy grails/ resonance
critical quality 151, 153
of wisdom = philosophy 92, 149
resonant association .... *see* music/
　positive power/ harmonic
　oscillation
Love
Brilliance/ *GUE*/ Unity/ Communion
　... 91

M

machine/ mechanism .... *see* also
　irreducible complexity/ minimal
　functionality/ teleology
biological organ/ system/ body . 10, 49,
　74, 84
conceptualised construction 135
functional design 74, 81
irreducible to physics and chemistry
　.. 74, 133
metaphysical/ Natural Dialectical
　machine 29
passively informed 130
whole greater than sum of parts 133
macro-evolution/ transformism 139,
　140
macro-evolution and innovation 139
Main Dialectic .... *see* dialectical factor/
　Primary Stack
man
as microcosm .... *see* cosmos
centaur 92
H. archetypalis/ memory man ... *see H.*
　archetypalis
H. electromagneticus .... *see H.*
　electromagneticus
H. sapiens .... *see H. sapiens*
metaphysical man/ memory man .. 100,
　see archetype/ caduceus/ mnemone
mass .... *see* matter/ energy/ Higgs
　mechanism
master routine ..... *see* code/ computation;
　top-down programming
material psychology ..... *see* neuroscience
materialisation
expression of (*tam* ↓) downward/
　gravitational vector .. 24, 33, 48, 61,
　76, 77, 118, 132
hierarchical order .... see fundamental
　order of creation/ creation/
　creativity/ cosmic fundamentals/
　hierarchy
informative/ energetic loss*see* entropy/
　gravity
of idea/ purpose/ design .... *see* act of
　creativity/ psychological force/
　teleology
orderly fall-out/ concretion/ fixation
　............. *see* exhaustion/ impotence/
　subtendence
materialism .... 115, *see* also faith
force and particles alone 33, 37
fundamental error 99
illusions of*see* illusion/ reality/ matter/
　consciousness
informal cults of matter - humanism/
　scientific atheism/ scientism87, 151
philosophy/ world-view ... 9, 11, 12, 97
materialism's immateriality .... *see*
　immateriality/ nothingness/ space-
　time
mathematics
archetypal geometry .... 106, *see* also
　vibration/ dot/ cosmic models
equation/ balance .19, 23, 117, *see* also
　balance/ equilibration/ *karma*
models nature 16, 117, *see* also cosmic
　models
numerical metaphor 116
strange equations 20

matter ....*see* also energy
  absolutely passive/ reflexively informed 69, 83, *see* archetype/ first cause physical, *see* archetype/ first cause physical
  aimless/ creatively impotent 65, 77
  can't create code ... 143, *see* innovation
  can't create mind/ information ....*see* mind/ information/ creativity
  condensed/ bulk matter = cosmic sink 16, 24, 42, 47, 50, 130
  cult of matter .... *see* materialism
  energetic entropy > sink *see* sink/ sub-state/ entropy/ exhausted phase
  eternal... 45, 127, *see* also steady-state/ multiverse
  illusory? 148
  made of energetic patterns 146
  matter-in-practice/ classical matter/ bulk form 35, 73
  matter-in-principle/ quantum forms 35, 72, 107, 128
  mind = matter/ brain .... *see* (neuro-)scientific reductionism/ neuro-hormonal system/ consciousness/ materialism
  no-life-in-it 42, 49, 59, 72, 88
  no-mind-in-it 10, 13, 49, 54, 60, 62, 69, 72, 100, 146
  non-conscious.. *see* non-consciousness
  physical essence .... *see* archetype/ potential matter/ potential (energetic-physical)
  potential matter....*see* potential matter/ archetype/ typical mnemone
  subjective annihilation/ sink 42, 74, 130
matter-in-practice ....*see* matter
matter-in-principle.....*see* matter/ energy/ force physical/ sub-atomic particle/ quantum aspects
mechanism .... *see* machine
meditation
  specific discipline of contemplation 154
memory ...... *see* also sub-consciousness/ mnemone/ archetype
  physical
    recording/ playback heads ....*see* brain/ engram/ hippocampus/ amygdala etc.
  psychological/ metaphysical idea registry 69, 79, 96
metabolism
  energetic operations
    photosynthesis.......................... 132
    respiration 132

metaphor ....*see* also cosmic models
  evolutionary 'as if' 117, 135, 154
  image/ metaphysical form/ symbol of the invisible................................. 18
  scientific metaphor .... 14, *see* also mathematics/ big-bang/ atomic/ evolutionary/ brain-as-computer/ scientific *inc.* cosmic models
metaphysic
  metaphysical egg 22, 69, 77, 119, 138, *see* also idea/ archetype/ first cause/ innovation/ conceptual development
  not constrained by physical law......55, 116
Metaphysical Egg ....*see* First Cause/ Archetype/ *Logos*
microcosm 36
mind
  = metaphysical entity .48, 52, 56, 60, 70
  consciousness
    = active information 73, 92
    animal perception .... 92, *see* also centaur
    human normality is triplex
      egotistical mind/ knowledge-to-fulfil-desire/ body-rationality ....92, *see* also animal perception
      higher mind/ knowledge-for-its-own-sake/ intellectual principal 92
      lower mind/ ignorance/ depression/ confusion/ negative intent/ evil-rationality 92
  informative element/ informant/ instructor 49
  loops
    lower physical/ mind-in-practice 92
    upper metaphysical/ mind-in-principle 92
  purposeful/ teleological force.....78
  subjective form/ image-generator/ formful consciousness 52
mind over matter 33, 36, 74, 107, 148, *see* also hierarchy/ cosmos/ conscio-material dipole
non-consciousness
  special case/ no-mind/ subjective absence/ impotence...... *see* non-consciousness/ impotence/ matter/ energy/ subtendence/ objectivity
sub-consciousness .... *see* also sub-consciousness/ mnemone/

190

archetype/ first cause physical
matrix of matter 128
universal mind 10, 35, 57, 60, 116,
117, 121, 123, *see* also implicit
universe/ archetype
minimal functionality 114
missing link
developmental/ anticipated target .. 143
nascent form 139
mnemone .... *see* also Glossary
memory/ psychological storage 57, 66,
73, 90, 93, 95, 116, 138
mnemonic control/ first cause physical
.... 100, 104
personal 90
sub-conscious
function = informant 90
psychosomatic mechanism 90
structure 101, 107
typical
archetypal 90, 96, 97, 99, 100, 101,
103, 104, 132, 137
morphogene 103
tripartite structure 101, 103
typical with every cell 101, *see* also
cell/ dormant in every cell
modular programming ...*see* code/ *top-
down* programming/ computation/
typology/ homology/ sub-routine/
analogy/ mosaic distribution
bio-module 136, 138
molecular/ cellular machinery. *see* also
cell/ bio-logic/ irreducible
complexity/ minimal functionality
energetic complements/ input-output
mitochondrion 103
enzymes/ metabolic pathways as bio-
machines 103, 108, 139
extra-cellular matrix (*ECM*) 103
structural complements/ 'flesh &
bones'
containers (e.g. membranes/
cytoplasmic gels).......... 102, 132
organisers (e.g. centrosome/
centriole/ cytoskeleton/ mitiotic
spindle) 103
morality
absolute/ relative 152, 154
chemical?
morality < genetic *DNA*? 37, 151
morality < molecular amorality?
.... 150
morality < neuro-hormones? 37
quality of intention 150, 151
morphogene
first cause physical/ formative
influence ....*see* potential matter/
archetype
influences every cell 101
logical/ conceptual bio-form 138
metaphysical blueprint................... 101
metaphysical software/ program 137
psycho-biological mechanism 103
psycho-biological/ archetypal
'intelligence' 101
psychosomatic mediator/ connector
.... 137
sub-conscious structure 101
sub-routine of typical mnemone ... 103,
132
morphogenesis
biological development 139, 143
resonant .*see* also resonance/ harmonic
oscillation/ Chladni
resonant modality 106
mosaic distribution ..... 138, 139, *see* also
Glossary/ convergence/ adaptive
potential/ homology/ variation-on-
theme
motion/ change .... *see* also space/ time/
relativity/ energy/ mind
(consciousness-in-motion)/ existence
(perpetually-in-motion)
< imbalance 122
> relativity 20
action-reaction .... *see* karma/ Law of
Motion
cyclical equilibration ...... *see* vibration
dialectical sense
'down' .... *see* gravity
'rest' .... *see* equilibrium/ balance
'up' .... *see* levity
fundamental nature of existence 18,
112, 147
information-in-orderly-motion/ music/
language 106
kinesis
anti-expression by (*tam* ↓) inertial
cosmic vector .... *see* gravity
deceleration 24
resistant inertia/ drag/ > static
fixity/ finish/ stop .... *see*
exhaustion/ negative power/
sink
special case/ no-motion/ rest.*see*
impotence/ negative
equilibrium
expression of (*raj* ↑) kinetic
cosmic vector .... *see* levity
acceleration & non-linear
motion 25
stimulus/ energetic input .... *see*
levity/ causal agent/ force
psychological or physical

unchanging velocity 25
expression of (*sat*) vectorless balance
  special case/ no-motion/ rest *see* source/ potential/ positive equilibrium
perpetual motion
  cosmic/ classical 20, 147
  kinetic theory of matter 147
  sub-microscopic/ quantum 147
Mount Universe .... *see* cosmic model
music .... *see* also harmonic oscillation/ resonance
  Master Analogy of Natural Dialectic 105
  natural music/ alphabet of notes ..... 84, 105, 125, *see* also alphabet/ cosmological language/ score/ Orpheus
mutation ... *see* also informative entropy/ *DNA* molecule/ gene/ innovation
  mutation > innovation? 139
  random mutation/ Darwinism's engine 133, 142, 149

N

Nanak 91, 153
Natural Dialectic 10, 14
Natural Dialectic of Polarity
  columnar structure .... 11, 54, *see* also Chapter 5 and dialectical stack
  Cosmic Dipole 47, 55
  Cosmic Infrastructure 15, 119
  dialectical factor. *see* dialectical factor
  Essential Monism and existential dualism 59
  Essential Principal 46
  existential principal 38, 54
  Master Analogy/ music 104, 105
  philosophical structure 7
  Primary Axiom .. 54, *see* also Glossary
  Trinity/ Triplex Nature. 17, 21, 23, 29, 61, *see* also cosmic fundamentals/ cosmos/ hierarchy
natural law .... *see* nature/ law/ order/ invariance/ conservation
natural selection/ Darwinism's regulator 139, 142
naturalistic methodology
  limitation 99
  materialistic assumption/ scientific method 12
nature
  holism
    physical cosmos = base-end of creation 55
  holism (including energetic physic)
    informative metaphysic is natural 140
  natural order/ fundamental order of creation ..... 15, 57, 117, 153, *see* also cosmic fundamentals
  naturalistic order/ laws of nature by chance?. 79, *see* also chance/ randomness
  psycho-physical nature 55, 151
  unnaturalism = purposeful design/ teleological engineering 117, 153
  naturalism (excluding informative metaphysic)
    immaterial = metaphysical = unnatural = super-natural 140
    natural sciences/ scientific naturalism < naturalistic methodology 55, 99
    naturalism = materialism ..... 69, *see* also neuro-scientific reductionism/ scientism/ faith
  nature = physical cosmos 55
negative power
  > materialisation/ confinement 57
  amoral expression of (*tam*) vector .. 76, 92
  involves informative/ energetic loss .... *see* entropy
  negative extremity 61
  selfish passion *opp*. love 92
  unintentional infliction of pain/ death .... 150, *see* also morality/ amorality
negentropy *see* also energy/ information/ order/ levity/ Glossary
  energetic
    expression of (*raj* ↑) fundamental 24
  informative
    teleological/ anti-randomising tendency 34, 56, 64, 78, 153
    psychologically/ metaphysically/ informatively dominant vector 56
neo-Darwinism.*see* index *STE/ GTE/* also glossary *PCM* and evolution
neuro-hormonal system
  alive as biochemicals 87
  chemical messenger/ informative hormone 102
  CNS (central nervous system)/ trunk route/ spine 107
  involuntary/ autonomic 107
  made of atoms/ molecules 87
  neuro-scientific imperative
    genes > nerves/ hormones > morality? 152
    nerves = thought/ memory?. 87, 95, 149, 152

systematic excretion of consciousness? 88
passively informative/ reflex 48
psycho-neuro-immunological system 107
psychosomatic mediator/ information exchanger 132
specifically codified 84
voluntary/ peripheral 107
neuroscience *see* neuro-hormonal system
neutrality..*see* also balance/ equilibrium
  detachment/ disinterest 21
  negative
    post-dynamic impotence 44
  positive
    fullness 21
    potential equilibrium 43, 44, 60
    pre-active potency/ poise 44
  Sat cosmic fundamental/ quality 21, 27
Newton Isaac 56
Nirvana ...... 91, *see* also Glossary
noise ...... *see* randomness
non-consciousness 58
  absolute ignorance/ perpetual oblivion 52, 61, 88
  body-state inc. *Homo sapiens*... 48, 81, 98
  cosmic base-level/ sub-state/ sink... 34, 55, 60, 88
  inanimate condition 147
  natural law = reflex behaviour 42
  objective being 62
  passive information 81
  physical phenomena/ forms 61, 66, 90, 97, 98
  psychological nadir *opp.*
    Psychological Zenith = Consciousness 88, 97
  psychology of physic 34, 118
  reflex/ automatic dimension 48, 75
  sixth state/ special case of consciousness (none) 94, 97
  subjective exhaustion/ void 61, 88
  subtendent pole 48, 88
  uncreative condition of matter/ material energy 13, 62, 67
Non-Existence *see* Essence/ Super-State/ Glossary
noology 86
Northrop Filmer 102
nothingness/ void
  absence/ anti-presence 55
  absolutely nothing 45, 55
  ex nihilo - from non-existence existence
  45, *see* also big bang/ zero-point
  information weighs nothing 65
  informative *and* energetic void/ void of voids/ impotent space 123
  informative potential 112, 116
  negative/ post-active exhaustion 44, 61, 123
  nothing-physical/ something-metaphysical ...... 9, 55, *see* also potential matter/ archetype
  positive/ pre-active potential 43, 44, 119, 121, 123
  unconscious oblivion 52
Nothingness/ Void
  Essence/ Potential 46
  Non-Existence/ Supreme Being 46
  Source 32, 104
  Void/ (N)One 40
NREM sleep 94
Nucleus ...... *see* Natural Nucleus/ Super-Nature

O

objectivity
  attempt to eliminate subjective context/ prejudice concerning observation 149, 153
  instrumental/ mathematical treatments of material phenomena 87, 150
  material condition/ objective being. 34, 55, 84, 94
  matter's special case of subjectivity (zero) 94
Objectivity ...... *see* Essence/ Absolute/ Consciousness
oblivion ...... *see* non-consciousness
opposite pole from Super-Consciousness 48, 65, 91
Om..53, 104, 128, *see* also *Logos*/ Sound and Light
Omega 127
opposites ..... *see* polarity/ dialectic stacks
order
  active order of mind 65, 78, 79
  archetypal ...... *see* archetype
  codified order ...... *see* code
  codified self-assembly by chance? 108, 137
  codified/ bio-logical self-organisation 145
  e principio/ from-principle-to-practice 15, 24, 46, 54, 57, 77, 105, 112, 113, 115, 118, 119, 121, 124, 132, 148, 151, 153
  fundamental order of creation ...15, 22, 29, 30, 46, 61, 73, 119
  informative communication ⬄ order 69, 78, 82, 104
  natural law/ reflex behaviour/

automatic order ...... 57, 69, 83, 117, 123, *see* also law/ archetype/ cosmo-logic/ nature
non-purposive complexity .. 69, 77, 82, 111
passive order of matter 65, 68, 77
psychosomatic order 100
purposive/ specified complexity 78, 81
schematic order .... *see* also conceptual development/ psychological development
social/ rules and regulations 151
top-down order . 34, 48, 69, 79, 83, 87, 118, 124, 140
uncodified/ physico-chemical self-organisation 112
organic development .... *see* biological development
oscillation .... *see* wave/ harmonic oscillation/ vibration

P

pain
< immorality 150
expression of constrictive (*tam*) vector 92
unintentional/ reflexive infliction .. 150
Paranada ..... *see* Logos/ Sound and Light
Pascal Blaise 77
passive complexity ..... *see* non-purposive complexity
passive information .... *see* information
passive mind .... *see* sub-consciousness
Pasteur Louis 142
peace/ rest .. 61, 91, 150, *see* also balance/ equilibrium
pendulum .... *see* vibration/ wave
Penrose Roger 113, 114, 116, 117
perfect mystic .... *see* saint
peripheral dialectic .... ***see*** **dialectical factor**
perpetual motion .... 26, *see* motion/ existence (mind-and-matter)/ atom
physical constants 13, 105, 116
physical objects/ events/ phenomena .. *see* energy/ matter/ polarity
physics 9, 22, 30, 61, 70, 84
pivot ... *see* point of balance/ centre-point
pixel/ quantum ..... 101, 102, 115, *see* also cell
Planck Max 116, 128
Plato 116
poise .... *see* equilibrium/ potential
polar coordinate .... *see* energy
polar dialectic .... *see* dialectical factor/ secondary stack

Polarity - Essential ..... *involves Primary Dialectic*:
Absolution ◇ relativity 57
Essence ◇ existence 28, 57
Neutrality
Balance ◇ levity/ gravity 24, 62
Neutrality ◇ + / - 60
Potential ◇ polar expression 44, 137
Sat ◇ *raj/ tam* 22, 24
Unity ◇ polarity/ duality 13, 23, 58, 60, 62, 137
Nothing ◇ something . 42, 44, 57, 123
One Truth/ anti-parallel world-views 12
Primary ◇ Secondary Dialectic 11, 112
polarity - existential *involves secondary dialectic*
bio-logical/ informed
neuro-hormonal 95
physical/ informed
large-scale/ matter-in-practice 24
quantum/ matter-in-principle 28
psycho-biological
active ◇ passive information ... 66, 79
archetype ◇ physical expression 97, 130, 137
immaterial ◇ material structure 106
informant ◇ informed 69
life ◇ non-life 61
mind ◇ body 61, 79, 134
perception ◇ action 32
psychosomasis 97
universal mind ◇ universal body 79
psycho-logical/ informative
contemplation/ sensation 90
mobile experience/ immobile memory 95
oscillations
asleep ◇ awake 91
politics external
market-place 150
strategic survival 150
Popp Fritz-Albert 102
Popper Karl 121
positive power
expression of (*sat*)/ (*raj*) cosmic fundamentals 76, 92
love *opp*. selfish passion ... 52, 92, 147, 148, 153
potential
energetic-physical
basic properties derived < field-in-

space 115
bio-logical 101
bio-nuclear potential/ *DNA* 69
electrochemical 102
material order/ archetype > matter-in-principle > matter-in-practice
 .... 35, 84, 101, 124
nuclear/ quantum expression of code/ blueprint 69, 83, 84
physico-chemical potentials ...... *see* Glossary
quantum agency/ matter-in-principle 43, 67, 69, 84

informative + energetic
(latent) field of possibilities 16, 44, 149
apparently formless/ unseen source 16, 17, 32
cosmic order/ potential > action > exhaustion 17, 18, 22, 30, 44, 61, 149
harmonic oscillation/ resonance/ music 105
hierarchically superior 73, 119, 125
informant (mind) > informed (physical pattern of behaviour)
 .... 49, 52, 73, 119, 149
point of balance/ poise ... 16, 18, 19
potential matter . 52, 55, 68, 72, 97, 116, 120, *see* also archetype/ transcendence/ first cause
physical/ implicit universe precondition/ prerequisite/ unexpressed pattern of behaviour 23, 95
pre-dynamic absence of motion 61, 79
principle/ intrinsic character/ quality 21, 23, 46, 54, 65, 92, 113, 119, 123, 124, 138, 148, 151, 153
source of typical action 55
unexpressed capacity 79
unrealised possibilities as *opp.* impotence (none) 114, 123

informative-metaphysical
codified arrangement/ blueprint stored/ memorised for later expression/ archetypal record 82
metaphysical egg/ idea 80
potential mind .... *see* Potential
purposeful scheme/ plan 23, 80, 137

Potential .... *see* Essence/ Neutrality/ Consciousness/ Source/ Archetype/ First Cause Psychological/ Super-State/ *Logos*

potential matter.. *see* potential/ energetic-physical
preordination .. *see* informative potential/ law/ order/ archetype
Primal Motion .... *see Logos*
Primaries
 Axiom of Materialism 64, 87
 Axiom of Natural Dialectic . 54, 64, 87
 Corollary of Materialism 9, 87
 Corollary of Natural Dialectic .. 11, *see* also Glossary
Dialectic 27
Primary Paradox 38

principle
metaphysical entity 119
principle > practice/ generality > individual cases... 112, *see* quantum aspects/ variation/ creation/ archetypal projection/ matter-in-principle/ matter-in-practice
probability .... *see* also chance/ potential/ variation-on-theme
0 = impossibility/ 1 = complete certainty 44, 114
collapse of wavefunction = 1 113
nature viewed statistically 114
Penrose computation 114
program .... *see* computation/ code/ teleological algorithm
promissory faith .... *see* faith
protein
 bio-electrical property .... 103, *see* also cell
proto-cell 144
PSI/ psychosomatic interface .... *see* psychosomasis
Psyche 86

psycho-logic
informative qualities 22
psychological development spiritual evolution 91, 154
psychological entropy/ information loss .... 93, *see* sleep/ ignorance/ randomness/ sub-consciousness
psychological force/ volitio-attraction ...... 68, 73, 75, 80, 92, 134, *see* also purpose/ will/ desire
psychological impotence .... *see* sub-consciousness
psychological negentropy > understanding/ learning/ lightness of being 92, 154
psychological record/ database .... *see* sub-consciousness/ memory

psychosomasis
archetypal memory/ program. 103, 104
border/ interface/ medium 13

informative transduction........ 104, 146
interface/ linkage.... 13, 48, 49, 66, 72, 79, 104, 107, 109
mind-matter exchange................... 106
morphogene..................................... 101
psycho-biological exchange 37, 49, 79, 107, *see* signal translation
sub-conscious/ psycho-biological mechanism. 48, 49, 79, 90, 97, 103, 107
Pythagoras................................ 106, 116

Q

quality ..... *see* also cosmic fundamentals
expression of cosmic fundamentals 21, 73
metaphysical value/ yardstick. 22, 149, 150, 151
quantum aspects
(*raj*) kinetic phase of material phenomena............ 35, 49, 104, 124
creation fields 125
determined quantum characters > natural law 112
HUP (uncertainty principle)........... 114
informant agency.............. 49, 103, 105
letters/ punctuation of cosmic language.................... 67, 69, 83, 84
psychosomatic agency?... 97, 107, 116
quantum theory...... 114, 121, 128, 129
sub-strate of potential matter 35, 73, 116, 118
super-strate of condensed/ classical matter...................................... 35, 73
whence quantum characters?........... 52

R

raj.........*see* cosmic fundamental/ positive influence
vector of levity 23
randomness 81
random variation 133
reality
exclusively materialistic/ scientific version 146
filtrate of brain............................... 146
hierarchical/ lesser realities/ relative appearances.................. 55, 147, 148
inclusive holistic (materialistic + immaterialistic) version 97, 147
material = energy/ immaterial = information 87
psychological reality 86, 115
quantum reality.............................. 113
special delusion
informative/ matter generates information............ 87, 115, 139

informative/ matter innovates... 139
neuro-scientific reductionism/ mind = matter 146
neuro-scientific/ mind = matter ..87
physical/ starry cosmos represents the whole truth..................... 148
Reality 148
reason
cosmos for no reason? 45, 112
gives meaning/ informs 80, 108
metaphysical exercise.... 117, 136, 153
no reason = irrationality................. 108
opp. randomness/ chance............... 137
orderly/ intellectual exercise............ 92
reason = cause 79
sub-reason/ whimsy/ thoughtless reflex/ instinct 92
relativity ....*see* also motion/ change
= motion/ change............................. 50
= relationship............. 13, 73, 119, 150
dialectical relativity 15, 24
Einstein's physical theories of....... 128
existential truth/ reality is relative ..20, 32, 54, 147, 148, 149
fundamental relativity....... 24, 105, 147
mind unifies/ orders relativities 116
moral 152, 154
opp. absolution 24, 62
religion/ formal canon
peripheral/ formal religion or world faith... 151
sub-religion 150
supra- or nuclear religion......... 91, 151
reproduction
> end-product/ reproductively enabled form 132
anticipatory/ teleological process ..132
asexual cellular/ cell cycle's reproductive phase 142
prokaryotic fissive sub-routines..... 142
resonance..*see* also harmonic oscillation
attunement 105, 107
electromagnetic 105
harmonic interaction...................... 105
quartz crystal 105
resonant association....... 104, 106, 107
synergetic 107
vibratory transfer of energy 105
retrotransposon ... *see* gene and Glossary, transposon

S

saint.. 149
Sat see cosmic fundamental/ tendency to balance/ neutralising influence/ potential
anti-vector/ quality of balance 23

scale .. *see* also conscio-material gradient
of being 40
of objectivity 111
of reality 148
of subjectivity 86
science of the soul 77, 154
secondary dialectic 28, 29, *see* dialectical factor/ secondary stack
seed .... 16, *see* also egg/ zygote/ idea
sensation .... *see* also neuro-hormonal system/ objectivity
opp. contemplation 90
outward diffusion of attention . 93, 146
physical connection 120
separation/ isolation/ differentiation ...*see* (dis-)unity
sex ...... 145, *see* also polarity/ symmetry/ reproduction
complex algorithm/ program .. *see* egg/ bio-logical development/ conceptual development
Shabda .... 53, *see* also *Logos*/ Sound and Light
Sherrington Charles 97
signal translation
mind <> body exchange 146
sub-routine of typical mnemone 103
sin/ swollen passion *opp*. beneficence . 92
sink
conscio-material base 15
extreme subtendency 61
impotence/ exhaustion/ end 34
psychological sink/ sub-conscious mind 41, 42
subjective annihilation 16, 42, 74
sub-state/ exhausted phase .. 17, 24, 36, 50
sleep 22, 48, 93, 94
snowflake 77
source
centrality 18, 34, 42, 112
dot/ point/ point of origin/ first hint of form/ primal motion 24, 123, 148
formless cause/ potential .. 17, 32, 123, 149
informative component 66, 80, 105, 119, 121, 122
inverted opposite of sink 16, 24, 26, 148
is matter source of mind? 121, 133
latent or incipient expression/ egg 121, *see* egg/ metaphysical egg/ idea
of creativity 11, 22, 92
of mind .... *see* Source
principal/ source of order . 46, 80, 105, 119, 123, *see* also order
projector .. 16, 18, 35, 45, 55, 111, 117

super-state/ potent start/ first phase 36, 148
space
archetypal property of pre-space .. 116, 117, 121, 123
dialectical trinity 123
formless page for matter's geometric script 69, 83
impotent/ exhausted sink 44, 123
inner space .... *see* psycho-space/ subjectivity
intra-atomic space/ vacuum 125
outer space/ range of extra-atomic vacuum densities 122, 125, *see* also void/ physical nothingness
physical nothingness/ emptiness/ abstraction 122, 123, 126
potential source/ vacuous quantum plenitude 44, 55, 123
pre-vacuous absence 55, 123
vacuum energy .... 122, 126, *see* ZPE
space-time
dynamic nothingness 45
materialism's immateriality/ apparent nothingness/ intangible physic .. 123
phenomenal extension/ basic physical coordinates 45, 124
species
artificial/ breeder speciation 139
stack .... *see* also dialectical stack/ dialectical factor
primary 27
secondary 28
standard model of particle physics 125
STE (special theory of evolution) .... *see* micro-evolution/ variation-on-theme)/ plasticity
Stern Otto 177
sub-atomic particle
collapse of wave function 114
electron/ electrical charge ... 49, 97, 98, 103, 105, 125
neutrino 125
neutron 125
photon
quantum of radiant energy ... 51, 97, 102, 103
proton 56, 122, 125
sub-consciousness .... *see* also mnemone/ archetype/ sleep
= memories/ mnemone/ psychological database ... 57, *see* also Chapters 15 and 16 *passim*
= psychological experience/ condition of sleep 93
archetypal memory 49, 83, 116
archetypal program 98, 100

archetypal/ psychosomatic control 102
automatic informant/ sub-conscious instructor 49, 103
dormant/ comatose conditions .. 48, 94, 102
dreaming & (semi-dormant) day-dreaming states 22, 93
finished state/ 'solid' mind 13, 90
fixed image/ frozen time 42, 79, 95, 99
immaterial element/ metaphysical entity/ form/ architecture 138
memories personal & typical .... *see* mnemone
non-waking zone 22, 72, 93
passive mind 68, 103
psychological context/ environment 13
psychological files/ 'photographic' records 57, 66, 67, 95, 99, 100, 138
psychosomatic 'sandwiched' zone . 49, 97, 107
rationally inaccessible 97
sub-state/ subtendent mind . 48, 52, 90, 119

subjectivity
experiential being/ awakened life 88
formful = mind 92
immaterial element 61, 75
inner subject as *opp*. outer object ... 61, 153
inward focus as *opp*. sensation/ physical action 79, 92
materialistic nervousness 86, 87
perspective involves informative meaning 22, 69, 150
range from 1 (Subjectivity) through mind to special case (0) matter .. 34, 37, 92, 146
special case (zero) = matter 16, 42, 62, 69, 74

Subjectivity .... *see* Essence/ Absolute/ Consciousness
sub-routine .. *see* code/ computation; *top-down* programming
sub-state ..... *see* impotence/ subtendence
(*tam*) position 35, 49
base pole/ physical universe 34, 49, 60
impotence *opp*. super-state potential 17, 19, 24, 36
informative/ energetic void 44
sink/ phase below 17, 47, 48
subtendence .... *see* also negative power/ sink
'dark/ nether pole' as *opp*. transcendence 47, 51
inferior effect/ next level down/ sub-state 47
non-consciousness .... *see* non-consciousness
opp. transcendence 61
Super-Consciousness*see* Consciousness
Super-Nature
(N)One 45
Consciousness . 151, *see* also Supreme Being/ Enlightenment
Natural Essence/ Nucleus/ Substance 88, 151
super-matter. *see* potential matter/ first cause physical/ archetype
super-natural = immaterial 48
super-state
(*sat*) position 34, 51
informative/ energetic poise 34, 51
source/ potential/ phase above .. 34, 51, *see* also potential/ informative component/ archetype
Super-State .... *see* Transcendent Information/ Source/ Potential
Supreme Being .... *see* Essence
survival .... *see* also reproduction/ reincarnation/ immortality/ Consciousness
> more life 108
cyclical equilibration/ homeostatic regulation ... 134, *see* also vibration/ homeostasis
economically comfortable 150
little deaths - conscious pain/ unconscious periods ..... 94, *see* also sleep/ pain/ negative power
non-survival/ death/ extinction 135, 139
self-seeking schemes 93, 152
what survives death? 101
switch
biochemical switches/ signal messengers 134
computational .... 136, 139, *see* also hierarchy/ computation/ circuitry/ teleology
genetic .... 101, 134, 136, 139, *see* circuitry/ computation
hierarchical switching system 138
homeostatic switches. *see* homeostasis
inter-modular/ between genetic routines .. 136, *see* also computation
inversion .... 61, *see* also inversion/ reflective asymmetry
symmetry ... *see* also Glossary and Index: balance/ equilibrium/ polarity/ mirror image
(*raj*)/ (*tam*) asymmetry-generators .. 24
(*sat*) quality of balance/ equilibrator/ symmetry generator 23
archetypal/ pre-actively latent 112,

113, 123
  breakage > variation-on-theme 112,
    see also variation
  central = neutral/ undifferentiated
    potential 104, 112
  geometry of balance/ proportion... 116,
    see also golden mean
  internal quantum/ gauge/ invariant/
    local/ space-time independent... 113
  invariance of natural law 112
  polar bifurcation 104, 112
  reflective/ polar asymmetry of
    complementary opposites 24, 57,
    60, 61, 112, 123
synchromesh 1
  cognizance <> personal mnemone ..95
synchromesh 2
  memory <> oblivion/ physical agency
    104
  mnemone <> body 104
  psycho-biological <> physical gearing
    104
  sub-conscious <> non-conscious
    gearing 104

T

tam .... see cosmic fundamental/ negative
    influence
  vector of gravity 24
Tao .... see dialectical operator
Tao (or Dao) 23
teleology
  causal/ provident phase 79
  conceptual development 77, 139
  fulfilment of expectation/ anticipation
    56
  goal-oriented program .... 135, 137, see
    also reason/ computation
  informative intent .... 64, see also force
    psychological
  study of purposive design in nature .56
  teleological opp. accidental 56
thought .... see active information/
    consciousness-in-motion/ mind
time
  era/ grade 124
  fundamental trinity 124
  species of time
    arrow/ straight-line/ DC-time 55
tonoscope 104
top-down 12
top-down programming 12
Tour James 142
transcendence .... see also super-state/
    potential
  'light/ upper pole' as *opp.* subtendence
    archetypal informant 73

next level up/ level above/ super-state
    51
  of logic/ mathematics 116
  source/ superior cause of inferior
    effect 24, 35, 51
  transcendent
    matter .... see potential matter/
      physical archetype/ physical
      absolute
    projection of physical phenomena
      . 45, 55, 111, see also first cause
      physical/ archetype/ source/
      informative potential
Transcendence
  metaphysical step from existence to
    Essence 53
transposon .... see gene and Glossary
tree of life
  family tree
    metaphorical/ Darwinian icon .140,
    144
Triangle of Truth 32
trinity .... see cosmic fundamentals/
    fundamental order of creation/
    hierarchy/ cosmos/ Natural Dialectic
  cosmo-logical
    macrocosmic see SAS figs. 1.3/ 2.3/
      2.6/ 3.1/ 13.1 also conscio-
      material (c-m) dipole/ cosmic
      models (ziggurat)
    microcosmic .... see figs. SAS
      13.1/14.1/ 15.1/ 17.3/ 17.4 also
      conscio-material (c-m) dipole
  dialectical 21
  physical/ informed .....see SAS figs. 3.3/
    10.2 also energetic trinity/ potential
    matter/ quantum matter-in-
    principle/ bulk matter-in-practice
  psychological/ informant .. 22, see SAS
    figs. 3.2/ 5.1 also informative
    trinity/ archetype/ Archetype
  psychosomatic/ psycho-biological ..see
    SAS figs. 15.2/ 15.4/ 15.5/ 16.1 also
    archetypal bio-classification
  reproductive....see SAS figs. 24.4/ 24.5
    also reproduction/ sex/ bio-logical
    development
  social ...see SAS fig. 26.4 also religion/
    politics/ law
truth
  graded/ sliding-scale/ hierarchical
    priorities 147, 149
  holism's = mind (based on
    consciousness) *and* matter 146
  materialism's = matter 146
  relative/ lesser existential truths....148,
    149

Truth 37, 91, 147, 148, 153

U

ultra-conscious = super-conscious 52
Uncreated Axis .... *see* also Essence/ Consciousness
understanding
   grasp of principle ...... 92, 118, 119, *see* also intelligence
   perception of cause/ how it works .. 77, 119, 128, 147, 153
unity
   (dis-)unity/ isolation (↓)/ apparently separate forms 147
   unification/ (↑) *raj* tendency 147
   unit/ integral whole e.g. proton/ atom/ protein/ cell/ multicellular body 147
   unitification/ (↓) *tam* tendency 147
   unity/ duality
        ◇ duality/ polarity .... 23, *see* also Natural Dialectic
        of complementary opposites 58
        union (↑)/ mergeance/ two-(or many)-into-one 147
Unity ...... *see* also Essence/ Communion/ Super-Nature
   Cosmic Monopole 58, 60
   Transcendent (N)One 40, 58
universal mind .... *see* mind

V

vacuum .... *see* space/ physical void/ nothingness/ ZPE
vacuum energy .... *see* vacuum/ levity
variation
   bio-logical
        evolutionary transformism 139, 141
        micro-evolution/ variation-on-theme 139, 140
   change/ relativity 77
   cosmo-logical
        low-level, local variation within high-level, general invariance 112, 114, 115
        variation-on-theme/ -principle/ -law.. 77, 82, 112, 114, 138, 139, 140
   on principle/ designed 78, 138

psychological invariant/ mind 88
special case - fixity/ invariance/ archetypal law 112, 114
the equivocation 141
vibration
   energetic oscillation. 24, 105, 128, 147
   harmonic 104, 105
   resonance/ resonant association 104, 105, 107
   wireless communication . 98, 104, 105, 106, 107, 132
virtuality .... *see* also ZPE
   archetype 69, 82, 113
   subliminal actuality 84, 122, 123
void .... *see* nothingness/ space
volitio-attractive/ mental push-pull ..... *see* psycho-logic

W

wave/ cyclical carriage of energy ..... *see* cosmic models/ vibration/ light/ harmonic oscillation
   sound/ light/ wave-borne information . 28, 53, 65, 69, 104, 105, 106, 114, *see* also harmonic oscillation
Wiener Norbert 65
will-power .... *see* force, psychological/ psycho-logic, psychological force
world-view
   needs account for metaphysic 10

Y

yin/ yang/ yin-yang ..... 23, *see* dialectical operator/ polarity

Z

zero-point *see* also balance/ nothingness/ apparent absolute/ ZPE
   complete exhaustion/ sink ... 42, 44, 50
   material absence 65
   non-existence 94
   pivot/ point of balance 24
   zero probability = impossibility 114
ZPE
   quantum energy of a vacuum ... 123
Zero-Point
   Nothingness/ Source/ Super-Nature. 41

The author has recently written a few more books (available from Amazon, Foyles, Waterstones, Barnes & Noble etc. and see website addresses on p.2):

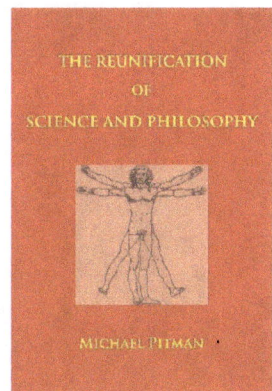
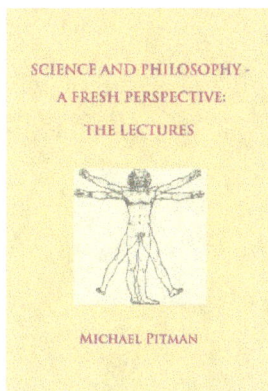
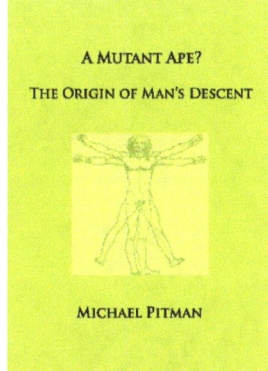
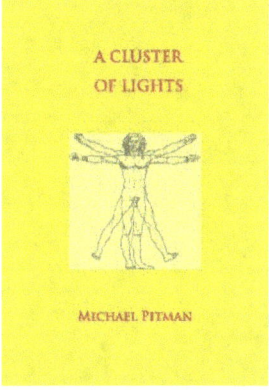
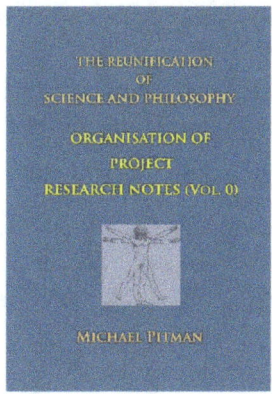

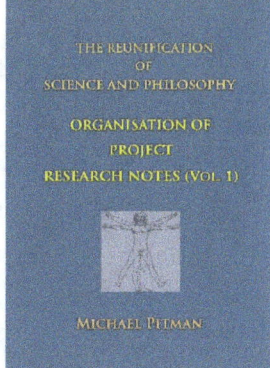

www.ingramcontent.com/pod-product-compliance
Lightning Source LLC
Chambersburg PA
CBHW072009070526
44583CB00015B/1395